Computational Intelligence and Human–Computer Interaction: Modern Methods and Applications

Computational Intelligence and Human–Computer Interaction: Modern Methods and Applications

Editors

Grigoreta-Sofia Cojocar
Adriana-Mihaela Guran

MDPI • Basel • Beijing • Wuhan • Barcelona • Belgrade • Manchester • Tokyo • Cluj • Tianjin

Editors
Grigoreta-Sofia Cojocar
Babeş-Bolyai University
Romania

Adriana-Mihaela Guran
Babeş-Bolyai University
Romania

Editorial Office
MDPI
St. Alban-Anlage 66
4052 Basel, Switzerland

This is a reprint of articles from the Special Issue published online in the open access journal *Mathematics* (ISSN 2227-7390) (available at: https://www.mdpi.com/si/mathematics/Computational_Intelligence_Human_Computer_Interaction).

For citation purposes, cite each article independently as indicated on the article page online and as indicated below:

LastName, A.A.; LastName, B.B.; LastName, C.C. Article Title. *Journal Name* **Year**, *Volume Number*, Page Range.

ISBN 978-3-0365-3937-9 (Hbk)
ISBN 978-3-0365-3938-6 (PDF)

© 2022 by the authors. Articles in this book are Open Access and distributed under the Creative Commons Attribution (CC BY) license, which allows users to download, copy and build upon published articles, as long as the author and publisher are properly credited, which ensures maximum dissemination and a wider impact of our publications.

The book as a whole is distributed by MDPI under the terms and conditions of the Creative Commons license CC BY-NC-ND.

Contents

About the Editors vii

Preface to "Computational Intelligence and Human–Computer Interaction: Modern Methods and Applications" ix

Anton Matveev, Olesia Makhnytkina, Yuri Matveev, Aleksei Svischev, Polina Korobova, Alexandr Rybin and Artem Akulov
Virtual Dialogue Assistant for Remote Exams
Reprinted from: *Mathematics* **2021**, *9*, 2229, doi:10.3390/math9182229 1

Emanuela Bran, Elena Bautu, Dragos Florin Sburlan, Crenguta Madalina Puchianu and Dorin Mircea Popovici
Ubiquitous Computing: Driving in the Intelligent Environment
Reprinted from: *Mathematics* **2021**, *9*, 2649, doi:10.3390/math9212649 17

Silviu Vert, Diana Andone, Andrei Ternauciuc, Vlad Mihaescu, Oana Rotaru, Muguras Mocofan, Ciprian Orhei and Radu Vasiu
User Evaluation of a Multi-Platform Digital Storytelling Concept for Cultural Heritage
Reprinted from: *Mathematics* **2021**, *9*, 2678, doi:10.3390/math9212678 41

Shih-Hung Yang, Yao-Mao Cheng, Jyun-We Huang and Yon-Ping Chen
RFaNet: Receptive Field-Aware Network with Finger Attention for Fingerspelling Recognition Using a Depth Sensor
Reprinted from: *Mathematics* **2021**, *9*, 2815, doi:10.3390/math9212815 65

Sergei Astapov, Aleksei Gusev, Marina Volkova, Aleksei Logunov, Valeriia Zaluskaia, Vlada Kapranova, Elena Timofeeva, Elena Evseeva, Vladimir Kabarov and Yuri Matveev
Application of Fusion of Various Spontaneous Speech Analytics Methods for Improving Far-Field Neural-Based Diarization
Reprinted from: *Mathematics* **2021**, *9*, 2998, doi:10.3390/math9232998 87

Gilberto Borrego, Samuel González-López and Ramón R. Palacio
Tags Recommender to Classify Architectural Knowledge Applying Language Models
Reprinted from: *Mathematics* **2022**, *10*, 446, doi:10.3390/math10030446 109

Adriana-Mihaela Guran, Grigoreta-Sofia Cojocar and Laura-Silvia Dioşan
The Next Generation of Edutainment Applications for Young Children—A Proposal
Reprinted from: *Mathematics* **2022**, *10*, 645, doi:10.3390/math10040645 135

Brigitte Breckner, Christian Săcărea and Raul-Robert Zavaczki
Improving User's Experience in Exploring Knowledge Structures: A Gamifying Approach
Reprinted from: *Mathematics* **2022**, *10*, 709, doi:10.3390/math10050709 151

Andrej Zgank
Influence of Highly Inflected Word Forms and Acoustic Background on the Robustness of Automatic Speech Recognition for Human–Computer Interaction
Reprinted from: *Mathematics* **2022**, *10*, 711, doi:10.3390/ math10050711 165

About the Editors

Grigoreta-Sofia Cojocar, Ph.D., is currently an Associate Professor at the Department of Computer Science at Babeș-Bolyai University, Cluj-Napoca, Romania. She received a Ph.D. in Computer Science in 2008. Her research interests are in the fields of software engineering, with a focus on aspect mining, programming paradigms, formal methods, and computational intelligence. Recently, she has been interested in how advances in artificial intelligence can be used to improve the human–computer experience of young and very young users.

Adriana-Mihaela Guran, Ph.D., is currently an Associate Professor at the Department of Computer Science at Babeș-Bolyai University, Cluj-Napoca, Romania. She received a Ph.D. in Computer Science in 2008. Her research interests are mainly focused on the design of interactive software systems using a user-centered approach. She is interested in applying participatory design with special user groups such as preschoolers. Combining HCI approaches with AI is one subject of interest in her recent research, and her main research focus is on using AI techniques to recognize emotions while interacting with edutainment and on adapting interaction based on the identified emotions.

Preface to "Computational Intelligence and Human–Computer Interaction: Modern Methods and Applications"

The domain of human–computer interaction has started to gain more and more attention due to advances in technological progress and the use of technology in almost every aspect of our lives. Empowering computers with human characteristics has long been a desire to facilitate more natural interaction between computers and their final users. The progress that has been made in the artificial intelligence domain has provided us with the opportunity to add more human capabilities to computers, such as user recognition, envisioning their actions and intentions, providing adapted support, or even identifying emotions and providing adapted feedback.

This book presents a collection of examples applying different artificial intelligence approaches to support better human–computer interaction. Some of the subjects discussed in the book are automatic speech recognition, processing, and analysis; fingerspelling recognition; intelligent driving; natural language processing; and smart edutainment.

The Guest Editors would like to thank all of the authors for their effort, dedication, and work in presenting how artificial intelligence and human–computer interaction can be combined to provide effective interactions through their papers. Their recognized expertise in the mentioned fields of artificial intelligence and human–computer interaction has contributed to the scientific quality of this book that we hope will be appreciated by readers.

Grigoreta-Sofia Cojocar and Adriana-Mihaela Guran
Editors

Article

Virtual Dialogue Assistant for Remote Exams

Anton Matveev *,†, Olesia Makhnytkina †, Yuri Matveev, Aleksei Svischev, Polina Korobova, Alexandr Rybin and Artem Akulov

Information Technologies and Programming Faculty, ITMO University, 197101 Saint Petersburg, Russia; makhnytkina@itmo.ru (O.M.); yunmatveev@itmo.ru (Y.M.); svischev@itmo.ru (A.S.); pikorobova@itmo.ru (P.K.); arybin@itmo.ru (A.R.); 287550@niuitmo.ru (A.A.)
* Correspondence: aymatveev@itmo.ru
† These authors contributed equally to this work.

Abstract: A Virtual Dialogue Assistant (VDA) is an automated system intended to provide support for conducting tests and examinations in the context of distant education platforms. Online Distance Learning (ODL) has proven to be a critical part of education systems across the world, particularly during the COVID-19 pandemic. While the core components of ODL are sufficiently researched and developed to become mainstream, there is still a demand for various aspects of traditional classroom learning to be implemented or improved to match the expectations for modern ODL systems. In this work, we take a look at the evaluation of students' performance. Various forms of testing are often present in ODL systems; however, modern Natural Language Processing (NLP) techniques provide new opportunities to improve this aspect of ODL. In this paper, we present an overview of VDA intended for integration with online education platforms to enhance the process of evaluation of students' performance. We propose an architecture of such a system, review challenges and solutions for building it, and present examples of solutions for several NLP problems and ways to integrate them into the system. The principal challenge for ODL is accessibility; therefore, proposing an enhancement for ODL systems, we formulate the problem from the point of view of a user interacting with it. In conclusion, we affirm that relying on the advancements in NLP and Machine Learning, the approach we suggest can provide an enhanced experience of evaluation of students' performance for modern ODL platforms.

Keywords: virtual dialogue assistant; natural language processing; machine learning; online distance learning

Citation: Matveev, A.; Makhnytkina, O.; Matveev, Y.; Svischev, A.; Korobova, P.; Rybin, A.; Akulov, A. Virtual Dialogue Assistant for Remote Exams. *Mathematics* **2021**, *9*, 2229. https://doi.org/10.3390/math9182229

Academic Editor: Grigoreta-Sofia Cojocar

Received: 31 July 2021
Accepted: 6 September 2021
Published: 10 September 2021

Publisher's Note: MDPI stays neutral with regard to jurisdictional claims in published maps and institutional affiliations.

Copyright: © 2021 by the authors. Licensee MDPI, Basel, Switzerland. This article is an open access article distributed under the terms and conditions of the Creative Commons Attribution (CC BY) license (https://creativecommons.org/licenses/by/4.0/).

1. Introduction

In this paper, we present a case for a Virtual Dialogue Assistant (VDA): an automated system intended to provide support for conducting tests and examinations in the context of distant education platforms. A *virtual dialogue assistant*, in this context and to the extent of our understanding, is not a common term; therefore, our approach for this presentation is first to introduce our model and its properties, define a case for it, and hypothesize about its utility. Then, we review our model in the context of the field of Natural Language Processing to draw similarities and distinctions with established problems researchers are working on, and suggest the role our model might play in the field. Finally, we explain our approach to various important problems we have to solve in order to, ultimately, assemble the model, and review our achievements and results at this point of our research.

The idea of distance education first appeared in the 18th century, and by the 20th century, there were schools focusing on it primarily [1]. The development of the World Wide Web in the 1990s revolutionized distance education, adapting it for the new reality of emerging post-bureaucratic organizations and globalization [2]. Since the beginning of the coronavirus pandemic in late 2019, according to the UNESCO Institute for Statistics data reports, more than 1.4 billion children in 186 countries were affected by school closures [3].

Worldwide, universities, especially the ones relying on revenue from international students, are facing major financial problems and are forced to either shut down or start offering their courses online [4]. However, it is not only educational institutions that have been online. According to reports, approximately 4% of corporations were using online learning in 1995, and presently, this number is at least 80% [5]. Various reports relying on data from major online education platforms, such as Coursera and EdX, present an evaluation of a compound annual growth rate for the digital education market at around 33% [6].

Our research targets some of the problems in the field of distance education, since we believe the data demonstrates that this field is rapidly growing and not only provides pure economic value, but also brings people convenience and the joy of learning, ultimately expanding opportunities for people to discover an interest in science and start performing their own scientific research, ultimately benefiting the community.

A significant impact of assessments on learning behavior was observed more than four decades ago, and even concepts like "assessment drives learning" were introduced [7]. Various forms of assessments were meticulously studied in the past few decades, and, it seems, there is no definite consensus on the exact relationship between forms of assessments and the quality of learning. Some researchers present cases that demonstrate a negative impact of assessment-driven education [8] and some point out that there is still much to be learned about the complex relationship between student learning and assessment [9].

However, even if someone has already formed an opinion on the role of assessments in the education process, online distance learning brings forward new circumstances that ought to be considered. Since distance learning brings an environment different from a traditional classroom, it is critical not to forget the essential roles of assessment: providing feedback to learners to remind them of what exactly they are doing and what they yet need to do to complete a course; offering a tool for evaluation of their progress; accumulating performance data to evaluate a course itself, if learners are performing poorly, there might be an issue with the course, not with the learners. It appears that overall, researchers seem to agree that various forms of assessment are necessary for distance learning, and new tools for assessment need to be developed, tested, and evaluated to support the modern learning environment [10,11].

It is also interesting to note that while the learning materials for a course might be universal for all students, the testing kit, in the ideal scenario, is different and adjusted for each student individually. In a one-to-one interview, a teacher is often equipped with past knowledge about a student and can manage the flow of the interview with the initial questions and the student's responses to previous questions. With distant examinations, there are several techniques, none of which seem to achieve the goal without significant drawbacks; for example, it is possible to prepare a large enough number of questions and present students with distinct subsets of questions that, however, does not mean the questions are adapted for each student, only that they are different; it is also possible to adjust selections of questions for a student with information about their intermediate indicators, if any, but even then, the possibility to tailor the following question based on the previous answers during the test is missing. It appears that no single solution is perfect; therefore, a compound approach is recommended [12].

Understanding that there is a demand for new tools for assessment in distance learning, in our research, we focus on investigating what features and qualities of such tools might be desirable and suggest an approach for designing, developing, and evaluating them. Specifically, our primary interest lies in the exponentially growing field of machine learning.

The field of Natural Language Processing (NLP) belongs to the intersection of linguistics and computer science. Beginning in the 1950s, the foundation of modern NLP was established by the end of the 20th century due to the significant increase in the amount of raw data and the development of traditional machine learning techniques. Over the past decade, modern machine learning techniques found their way into NLP. For our research, we do not limit ourselves to particular methods or algorithms, but instead borrow from the rich history of the development of the field, noticing the advantages and disadvantages of various approaches and techniques.

The development of NLP begins with symbolic algorithms [13]: methods that use rules and grammars to parse and model syntactic structures to imitate how humans apply rules and grammar to sentence recognition. With the development of machine learning techniques, they found their way into NLP; the emergence of significantly large textual corpora allowed for the application of statistical methods to extract information, discover patterns, and predict missing data, that is, to build probabilistic models [13]. It was noted that, in general, symbolic techniques offer more compact and high-level representations than statistical models, but lack in the level of robustness and geerality [14]. In the last decade, the development of deep learning models relying on increasing computational resources (specifically, Graphical Processing Units) and parallelization, allowing for building artificial neural networks with billions of trainable parameters, transformed many fields of study and, along with even further increasing amounts of available raw and annotated data, made it possible to solve many complex problems using *brute force*; naturally, deep learning techniques are now widely applied to NLP problems as well [15]. It is not always the case that to solve a problem, it is necessary to employ the most powerful tool; often, a weaker tool might be more efficient. If a tool provides a satisfying outcome for a task but requires less computational resources or research and development time, it might be reasonable to utilize it over a more powerful tool. With that in mind, in our work, we attempt to avoid excessively narrowing down the list of techniques to consider and focus on exploring a wider range of methods, since our primary goal is not to achieve the best outcome for each specific problem but, instead, to demonstrate the potential power of a complex system that utilizes sufficiently accurate solutions to the specific problems and, by achieving synergy between them, produces a higher-level outcome.

Broadly speaking, NLP techniques might be categorized into three groups. Initially, the idea was to operate on manually constructed knowledge bases and hand-coded symbolic grammars or rules [16]. In the 1950s, researchers working on developing fully automatic systems for machine translation—one of the core problems in NLP—were highly optimistic about advances in formal linguistics and computational power and were expecting the problem to be effectively solved in less than a decade. Soon, however, the realization that the problem might be more complex than initially estimated occurred, with prominent researchers even expressing an opinion that the problem might be unsolvable in principle [17]. It appears that a rule-based approach might be appropriate and efficient for narrow situations but requires an exceeding amount of manual input for more general problems. In this work, we take advantage of this property of rule-based methods and apply them when we work with a problem, for which it is more important to demonstrate a proof-of-concept rather than the best possible accuracy.

Another category of NLP techniques is empirical for statistical methods. Compared to rule-based methods, the main difference is that the manual input is replaced with a combination of training data and a learning system that operates on the training data to produce a knowledge base. A knowledge base is a general term for various data structures that might be appropriate for particular problems [16], for example, ontologies. Some authors note that ontologies, even as a specific kind of knowledge base, might be too broad of a concept to specify their role explicitly; however, it is relatively clear that in virtually any case, they provide, at least, access to the meaning of terms, concepts, or relations [18]. In this work, we rely on the idea of ontologies for providing us with a tool for expressing the goal, the problem, and the reference solution. We expand on that further.

The state-of-the-art category of NLP techniques results from reaching the critical point in terms of computational resources and volumes of raw and annotated data in the past decade. For example, deep learning models approach hundreds of billions of parameters and terabytes of the total model size in a modern way [19]. Deep learning models already demonstrated exceptional performance for computer vision [20] and speech recognition problems, and more fascinating results appear to be on the way. The most common architectures for deep learning models include recurrent neural networks (RNNs), convolutional neural networks (CNNs), recursive neural networks, and generative adversarial networks (GANs) [21]. Across various deep learning models, several concepts are proven to be core for most NLP problems.

The first is feature representations or embeddings. The aim of feature embedding is to encode text input into a form that focuses on highlighting particular characteristics of the input that are important to a specific problem. Depending on the input and the type of problem, embedding can be performed at the level of single characters, words, phrases, and even paragraphs.

The second concept is sequence-to-sequence modeling. In many problems related to human intelligence, data points are not independently and identically distributed, but instead depend on the preceding or following data points, and language is a straightforward example of sequential information. Moreover, for many NLP problems, not only is the input often sequential, but the output is also expected to have a similar nature, for example, in machine translation. One common approach to sequence-to-sequence modeling is to employ an encoder that consumes input and produces an intermediate output, and a decoder that transforms that output into the desired form.

Another important concept is reinforcement learning. There are several obstacles to sequence-to-sequence modeling, such as exposure bias and inconsistency between the train/test measurement; one approach to handle these issues is to apply reinforcement learning techniques and enhance a model to only rely on its own output, rather than the ground truth, and to directly optimize the model using the evaluation measure [22]. Training deep learning models for NLP problems requires significant computational and storage resources and is often performed by distributed networks of GPU servers [23]. For the purposes of our research, we do not aim to modify or reproduce any particular deep learning model; while it might be helpful to develop and train deep learning models to precisely fit to the requirements of our tasks, we believe, in our particular case, it is neither efficient nor necessary.

Among the three approaches, deep learning appears to have one clear advantage, and that is virtually unlimited scalability. Even a decade ago, many problems were deemed unapproachable by computers, for example, the game of Go, but recently the game was, essentially, solved with deep learning [24]. It appears all of the modern NLP problems can be, eventually, solved with deep learning [21]. However, at this moment, we believe it is not practical to simply delegate any problem to deep learning algorithms: the field of artificial intelligence is still relatively new, and most of the significant practical results were only produced in recent years. In our research, instead, we consider all three approaches and, if anything can be recognized as the key idea, it is the idea of ontologies. While deep learning solutions are powerful tools for achieving results, they are still mostly black boxes that provide little insight into the problems; on the other hand, while the idea of ontologies in itself is rather abstract, it provides an insight into how a solution to a problem might be constructed.

Originating from philosophy, the concept of ontology was adopted by computer science, particularly Artificial Intelligence and Computational Linguistics, either to refer to general ideas of conceptual analysis or domain modeling, or to describe a distinct methodology or architecture of a specific solution for data collection (mining), organization, and access [25]. Guizzardi [26] defends the notion that "the opposite of ontology is not non-ontology, but just bad ontology". This notion is vital to us, since it allows us to employ the concept of ontology as a basis for our reasoning about the fundamental nature of the problem we consider and solutions to it; in other words, we believe it is sufficient to demonstrate that if there exists a solution that shows acceptable performance with respect to an incomplete or imperfect ontology, then this solution has to be considered a candidate for the ultimate solution to the problem, given our transcendent goal of building "good" ontologies.

One example of this ontology-based approach is the Wolfram System. The web page for the Wolfram Data Framework says [27]:

As by far the largest computable knowledge system ever built, Wolfram Alpha has been in a unique position to construct and test a broad ontology.

For example, met with a query "how many goats in Spain", the system produces [28] an interpretation of the query, a graph with livestock population historical data, and, ultimately, the result (3.09 million as of 2016). While the underlying knowledge base (ontology) of the system is not yet "perfect" and is constantly updated via data mining, it presents a valuable example of the capability and practicality of the ontology-based approach. This example is also important to us, since it is closely related to the fundamental problems we consider.

A virtual dialogue assistant for remote exams has at least two core features: question generation, and evaluation of students' answers. In the literature, those problems are approached in various ways.

There are methodological recommendations for composing exams and tests, including suggestions for managing the performance evaluation with automated systems. It is recommended to include an assortment of question types in an increasing difficulty progression:

1. Open question
   (a) Fill-in-the-blank question
   (b) Subjective question
2. Objective Test Question (OTQ)
   (a) True/False questions (statements that can be either true or false)
   (b) Closed or multiple-choice question (MCQ)
   (c) Sequencing questions (require sorting a set of items by some principle)
   (d) Matching questions (require interconnection of corresponding elements in two given sets)

Fill-in-the-blank questions are, generally, more straightforward to generate automatically than wh-questions. Fill-in-the-blank questions can be categorized as either a closed type [29,30] when there are candidates present, or an open type when they are not [31].

One obstacle for closed-type fill-in-the-blank question generation is the selection of reasonable but wrong potential answers.

At least in the past decade, there were multiple attempts to approach the closed question generation problem, resulting in a commonly acknowledged solution [32]. Modern research in this field primarily focuses on improving specific steps of the algorithm. The algorithm takes a text fragment as an input: usually, a segment of learning materials or specialized text. Some implementations of the algorithm work with formalized text structures, but, in general, the algorithm is supposed to work with arbitrary texts.

The first step of the algorithms is preprocessing of the input to obtain features that will be used at the next step. Two sources of input are considered here: (1) text in a natural language and (2) a structured representation of knowledge in a field with preset categories and relationships, for example, ontologies [30]. For texts in a natural language, the most common embeddings are statistical features (tf-idf), semantic embeddings, syntactic parsing, and POS-tagging.

The second step is the selection of a sentence that would be a source for a generated question. There are algorithms that can use a pattern to search for such sentences. A pattern might be a specific sequence of parts of speech. Another approach is machine learning, for example, summarizing [33]. The third step is key selection. The most common techniques are selection by word frequencies, pattern search, and standard machine learning models. The fourth step is question generation. The options are: to rephrase the sentence, to construct a question by applying a pattern, or to keep the original sentence but remove the key word. The fifth step is the selection of possible answer alternatives. The choice might be based on an ontology, synonyms, or nearest neighbors in the semantic space. The final step is post-processing: additional operations required to fit the generated question to the requirements.

Modern implementations are primarily based on heuristics and patterns, which make them more stable and easier to interpret. On the downside, the large number of parameters makes it considerably challenging to set up. Using patterns for question generation and key word selection, on the other hand, limits the range of possible questions that might be generated.

For open-question generation—questions without a predetermined set of possible answers—the most common approaches are: (1) the open-closed question generation, which is a type of fill-in-the-blank question; and (2) generation of a question employing auxiliary phrases, such as "What is ...", "What constitutes ...", and so forth.

Answers to objective questions can be clearly interpreted and objectively evaluated as either correct or incorrect [34].

However, the development of neural network solutions based on transformers and the emergence of GPT-3 and BERT language models for the Russian language and mT5 multi-lingual model demonstrates the possibility of improvement for question-and-answer systems. One of the remaining issues is the processing of text segments with mathematical symbols. There are few papers on the topic of Mathematical Language Processing (MLP), and the research on the question generation for such text segments in the Russian language is yet to emerge. It appears that learning materials, particularly in STEM, rather often contain mathematical symbols and equations, meaning the solution to the problem would be desirable.

Answer evaluation is a critical part of the remote examination. The most complicated form of evaluation is the evaluation of answers to open questions. In a classroom, the examiner assesses how close the answer is to the correct answer, and one important criterion is how close the keywords are from the answer to the keywords of the correct answer. By utilizing vector representations of words in a vector space with semantic differences between words as a factor in the distance measure, it is possible to construct a system that can handle the diversity of a natural language and variability in phrasing. In this work, we are exploring algorithms for comparing dependency grammars employing vector representations of words.

In the context of NLP, relevance is a criterion used to evaluate the performance of a search engine; in other words, how close the output of the system is to what was prompted. This concept can be applied to question-and-answer and remote examination systems. For the document ranking problem, it is sufficient to have only a relevance measure; however, when there is only one document (the student's answer), it is necessary to establish a threshold to categorize the answer as either relevant or not relevant to the correct answer. For comparing a student's answer and the correct answer, we experiment with dependency grammars [35].

One relatively simple way to compare dependency grammars [36] is to count intersections of sub-trees representing semantic relationships; then the similarity between two text segments can be calculated as

$$E = \frac{|Q \cap T|}{|Q|}, \qquad (1)$$

where E is similarity, Q is the set of semantic relationship tuples for the correct answer, and T is the set of semantic relationship tuples for the student's answer.

A more complicated approach is predicate matching [37]. In this case, it is not only the pairs of nodes, but all semantic relationships (so-called predicate relationships) at the verb that are considered.

Another method is fuzzy depth-first search [38]. This method also relies on dependency trees and can be summarized as traversing the trees simultaneously and adding or removing penalties depending on the item type and value.

Mathematical Language Processing is a field on the intersection of computer linguistic, formal languages theory, and, recently, machine learning and artificial intelligence. Mathematical language is defined as any expression with digits, variables, or operators. Unlike natural language, mathematical language is consciously designed, since it was created by people. This field shares similar goals with NLP, and the main topics are:

1. Search for similar expressions;
2. Extraction of structural information, for example, variable names;
3. Transformation of expressions between different forms, for example, LaTeX and MathML.

Unlike NLP, MLP is at the early stages as a separate field of study. While machine learning techniques are already established for many NLP problems, the application of them to MLP is still not widely researched [39]. Some authors note that direct application of the NLP machine learning techniques does not produce results of similar quality [40]. The existing methods for feature extraction can be adopted for mathematical expression by applying them to the level of symbols; in other words, by considering an expression as a text segment and a mathematical symbol as a word (token) in terms of NLP. In this way, techniques for vector representation, distributed semantics, and universal language models can all be applied to MLP.

There are examples of adopting the TF-IDF method to MLP [39]. The research was conducted with classification and clustering problems for texts with mathematical expressions. Two datasets were used for training and testing: SigMathLing arXMLiv-08-2018 and NTCIR-11/12 MathIR arXiv. The advantages of this approach are:

1. Simple implementation;
2. Fixed length feature vector;
3. Relatively small text corpus for training.

The disadvantages are:

1. The context of an mathematical expression is not considered;
2. Insufficient representation.

In our approach to the problem, specifically, our focus on online education platforms, there is another interesting aspect to consider. Among various ODL platforms, there is a typical structure to the organization of learning materials or courses. Here, we are interested in one specific trait: a comment section. While details may vary between platforms, we generalize a model of an ODL platform and suppose that a course is split into sections in some sequential order, each containing a segment of text representing learning materials for that section of the course. We assume the students subscribing to a course have access to comment sections specific to the sections of the course. Here, we are interested in one specific feature of such a model: the emotional reaction of the students represented by the comments they submit. We do not assume whether a positive or a negative response to a segment of learning materials correlates to the quality of the materials; we are simply interested in investigating whether or not the presence of a reaction, followed by an adjustment to the question generation part of the system, may affect the performance evaluation of students. The methods and techniques for emotion detection that we investigated in one of our papers are related to this research [41].

One related topic that, however, is beyond the scope of this paper is audio language processing. Notably, we want to mention on-device speech recognition [42] for the following reason: while traditionally, speech recognition is a resource-intensive task that virtually required a client-server implementation similar to Apple's Siri, modern techniques allow for on-device implementations, which not as much implies technical details, but presents a different view on architecture of solutions employing speech recognition as part of a pipeline; therefore, it allows us not to consider it as a task requiring specific accommodation, but simply as an ad hoc utility.

2. Methods

2.1. Semantic Relationships Extraction

Extraction of semantic relationships from unstructured text is essential for building knowledge bases, thesauruses, ontologies, and information search. Semantic relationships, also called paradigmatic semantic relationships or, less often, lexical relationships, are relationships between lexical entities (words, phrases) within a restricted topic. Semantic relationships specifically describe connections and differences between words and phrases. There are several common types of relationships: synonyms, type-of relationships (hyponyms and hypernyms), part-of relationships (meronymy and holonymy), antonymy, and converses (relational antonyms).

Most often, ontologies, regardless of the language, are comprised of several entities: instances (concrete, such as people, buildings, etc., or abstract, such as numbers, feelings, etc.), concepts (groups or classes of objects), attributes (instance characteristics that have a name and a value for storing information), relationships and functions (describe dependencies between entities in the ontology), and axioms which represent restrictions.

For this research, we employ a dataset in the Russian language and experiment with extracting semantic relationships via rule-based and machine learning-based methods from it.

We picked the dataset RuSERRC described in Bruches et al. [43]. This dataset is a collection of abstracts to scientific papers in information technologies with 1600 annotated documents and 80 manually annotated with semantic relationships: CAUSE (X yields Y), COMPARE (X is compared to Y), ISA (X is a Y), PARTOF (X is a part of Y), SYNONYMS (X is a synonym of Y), USAGE (X is used for Y). For our purposes, we select the ISA, PARTOF, SYNONYMS, and USAGE relationships; however, we also enhance it with HYPON (hyponymy and hypernymy) and TERM (a term and its definition) semantic relationships. Examples of these records are shown in Table 1.

The total number of records for semantic relationships is 599: 99 for ISA in the original dataset and 14 added by us, 90 for PARTOF and 8 added, 33 for SYNONYMS and 13 added, 311 USAGE and 14 added; HYPON and TERM relationships were not present in the original dataset, and we added 15 and 51 of them, respectively.

Table 1. Examples of semantic relationship records.

<e1>Естественный язык</e1>(ЕЯ) - <e2>язык, используемый для общения людей и не созданный целенаправленно</e2>. <e1>A natural language</e1>(NL) is <e2>a language which is used for people to communicate and which is not consciously designed</e2>.	TERM
Примерами <e1>естественных языков</e1>являются <e2>русский</e2>, <e2>английский</e2>, <e2>китайский</e2>, <e2>казахский</e2>др. Some examples of <e1>natural languages</e1>are <e2>Russian</e2>, <e2>English</e2>, <e2>Chinese</e2>, <e2>Kazakh</e2>, etc.	ISA
К <e1>формальным языкам</e1>относят <e2>язык математической логики</e2>; <e2>языки программирования</e2>; <e2>языки, порожденные регулярными выражениями</e2>, <e2>языки конечными автоматики</e2>, <e2>грамматикой Хомского</e2>и др. Some examples of <e1>formal languages</e1>are <e2>the language of mathematical logic</e2>; <e2>programming languages</e2>; <e2>languages emergent from regular expressions</e2>, <e2>closed-loop automations</e2>, <e2>Chomsky grammars</e2>, etc.	HYPON
<e1>Языки, созданные целенаправленно</e1>, называют <e2>искусственными языками</e2>. <e1>Consciously designed languages</e1> are called <e2>artificial languages</e2>.	TERM

Table 1. *Cont.*

</e1>Лексический анализ</e1>(<e2>токенизация</e2>) - выделение в тексте слов, цифровых комплексов, знаков препинания, формул, и т.д. *</e1>Lexical analysis</e1>(<e2>tokenization</e2>) is a process of selecting words, digits, punctuation marks, equation, etc. from a text.*	SYNONYMS
При аналитическом выражении грамматических значений <e1>слова</e1>типично состоят из <e2>малого числа морфем</e2>, при синтетическом - из нескольких. *From the point of view of analytical expression of grammatical meaning, <e1>words</e1>typically consist of a <e2>small number of morphemes</e2>, but from a greater number of morphemes from a synthetic point of view.*	PART_OF
<e2>Разработка информационной системы</e2>по клещевой опасности на основе <e1>онтологии</e1>предметной области Предложен подход к разработке состава и структуры Интернет - ресурса на основе онтологии предметной области. *<e2>For development of an information system</e2>for acari ticks danger prevention based on <e1>ontologies</e1>, an approach to development and organization of a web-portal was suggested.*	USAGE

2.1.1. Rule-Based Approach

The sentiment extraction begins with preprocessing of source texts, their annotation, and a compilation of rule sets; each rule is then applied to each sentence, and, finally, the extracted semantic relationships are evaluated. When the recall for a rule is below 0.7, the rule is removed from the set. Via this process, for semantic relationships extraction, 84 rules were selected (see Table 2). Since the rules are fine-tuned to the dataset if new texts are added to the collection, the rules have to be reevaluated.

Table 2. Examples of the selected rules (for demonstration purposes, verbs are translated from Russian into English, while the original structure and punctuation are kept intact).

	Examples of Rules	
Terms	$TERM—is $* $TERM, $*,—is $* $TERM,—is $* $TERM—is $* $TERM—$* $TERM—$* $* $TERM—$* $* ($TERM—$* $TERM is called $* $* called $TERM $TERM means $* $TERM solves $*	$* contraposed $TERM $TERM performs $* $TERM is defined $* $TERM marks $* $* $TERM means $* $TERM means $* $TERM is $* $TERM , $*, is $* $TERM is expressed by $* $TERM states $* $TERM (is performed by) $*

One necessary step for information extraction is preprocessing. Here, we execute a typical pipeline: stop-word removal, tokenization, lemmatization, and, optionally, correction. For performance evaluation, we calculate the F-score (Table 3).

Table 3. Performance of the rule-based approach.

	PART_OF	ISA	USAGE	HYPON	SYNONYMS	TERM
F-score	0.31	0.46	0.52	0.41	0.35	0.65
Precision	0.37	0.31	0.50	0.28	0.22	0.48
Recall	0.27	0.94	0.54	0.70	0.80	0.94

2.1.2. Machine Learning-Based Approach

In this work, we experimented with a neural network architecture from Bruches et al. [43]. This neural network has four layers. The first layer processes features from the pre-trained model Ru-Bert, where this model for the Russian language was trained on texts from the Russian segment of Wikipedia and news articles from the website, Lenta.ru. The volume of the pre-processed dataset is close to 6.5 gigabytes, 80% of it from Wikipedia. The second and the third steps apply a mask obtained from data annotations in such a way as to classify tokens into semantic relationship categories. The final layer outputs semantic relationship labels, and, finally, the data are passed through softmax. The training was performed with 10 and 20 epochs with the Adam optimizer and a starting learning rate of 0.0001.

The performance of the model with 20 epochs and a 0.0001 learning rate for each type of semantic relationship is shown in Table 4.

Table 4. Performance of the machine learning-based approach for each type of semantic relationship.

	PART_OF	ISA	USAGE	HYPON	SYNONYMS	TERM
F-score	0.78	0.95	1.00	0.86	0.86	0.29
Precision	0.78	1.00	1.00	0.81	0.81	0.33
Recall	0.78	0.90	1.00	0.91	0.91	0.25

2.1.3. Example of Question Generation

The semantic relationships extracted via both of the described methods can then be used to generate questions (Table 5).

Table 5. Examples of questions generated via extracted semantic relationships.

Question Pattern	Question in Natural Language
What is $TERM?	Что такое естественный язык?
Which kinds of $TERM there are?	Какие есть виды языков?
What are some examples of $TERM?	Какие есть примеры эльфийских языков?
Which parts $TERM consists of?	Что является частями анализа языка?
What is $TERM a part of?	Частью чего является Синтез языка?

For alternative (incorrect) answers, we used parts of other records of the same semantic relationship type. Here is a full example of a test constructed via question generation based on a text segment:

Естественный язык (ЕЯ)—язык, используемый для общения людей и не созданный целенаправленно. Примерами естественных языков являются русский, английский, китайский, казахский и др. Языки, созданные целенаправленно, называют искусственными языками. На данный момент их уже больше 1000, и постоянно создаются новые. Обработка естественного языка (Natural Language Processing, NLP)—общее направление искуственного интеллекта и математической лингвистики. Оно изучает проблемы компьютерного анализа и синтеза естественных языков.

Вопрос к отрывку №1:

Что такое естественный язык?

1. Язык, используемый для общения людей и не созданный целенаправленно. 2. какое-любо конкретное значение, которое может принимать данный признак (ключ). 3. лингвистическеие процессоры, которые друг за другом обработывают входной текст.

Что такое искуственный язык?

1. язык, созданный целенаправленно, 2. удаление значительной части "морфолонгического шума" и омонимичности словоформ 3. явление, при котором синтаксические конструкции имеют близкие значения и способны в определенных контекстах заменять друг друга.

Natural language (NL) is a language used to communicate between people and is not purposefully created. Examples of natural languages are Russian, English, Chinese, Kazakh, and so forth. Languages created purposefully are called artificial languages. At the moment, there are already more than 1000 of them, and new ones are constantly being created. Natural Language Processing (NLP) is a general area of artificial intelligence and mathematical linguistics. It studies the problems of computer analysis and synthesis of natural languages.

Question for excerpt 1:
What is natural language?
1. The language used to communicate between people and not purposefully created;
2. Any specific value that a given attribute (key) can take;
3. Linguistic processors, which one after another process the input text.
What is artificial language?
1. Language created on purpose;
2. Removal of a significant part of the "morpholongic noise" and homonymy word forms;
3. A phenomenon in which syntactic constructions have close meanings and are capable of replacing each other in certain contexts.

2.2. Deep Learning-Based Question Generation

For this problem, we experiment with neural network architectures based on transformers for three tasks: key selection, question generation, and selection between possible answers. The workflow is the following: a question is generated based on a text fragment, where this question and the text fragment are then used to generate an answer, and finally, the text fragment, the question, and the answer are used to generate alternative answers. We choose the pre-trained models ruGPT and mT5, and we pick the smallest variations (2 and 4 gigabytes, respectively) since we are primarily interested in proving the concept rather than achieving the highest possible precision. mT5 is a model often employed for seq2seq problems, and it contains an encoder that translates the input text into a latent vector space, and a decoder that takes the output of the encoder and its own output from the previous pass to produce a new output. ru-GPT-3 is a language model that only employs the decoder part of the original transformer architecture. Since the question generation from a context and the generation of an answer from a context and a question are sqe2sqe problems, we expect mT5 to produce more accurate output; however, we notice that question generation with ruGPT-3 produces a more stable and coherent result. For fine-tuning, we selected the RuBQ (3000 records) and SberQuAD (50,000 records) datasets, and for evaluation, we calculate perplexity (Table 6).

It is interesting to notice that an increase in batch size improves the performance. Since ruGPT-3 appears to be a better fit for our problem, next, we adapt it via zero-shot training: feeding a generative model with some patterns for it to fulfill it with actual data.

Table 6. Performance of the mT5 and ruGPT-3 models.

Model	Problem	Perplexity (Training Set)	Perplexity (Test Set)	Batch Size	Iterations	Dataset
ruGPT-3	Question	1049	2657	4	2959	RuBQ
ruGPT-3	Answer	1356	316	2	5917	RuBQ
mT5	Question	1357	1176	1	11,847	RuBQ
ruGPT-3	Question	157,259	102,679	1	85,328	SberQuAD
ruGPT-3	Question	6113	-	2	2096	SberQuAD
ruGPT-3	Question	2491	-	4	1048	SberQuAD

We experiment with the *Large* version of the model with 760 million parameters trained by a team from Sberbank with the supercomputer Christofari. Over 600 gigabytes of data (mostly in the Russian language) was used for training the model: literary fiction, Russian and English segments of Wikipedia, news web-portals, public categories of Pikabu.ru, a collection of articles for a popular science web-portal 22century.ru, materials from banki.ru, Omnia Russica text corpus, snapshots of GitHub and StackOverflow. The ratio between data in Russian and English used by the authors for training the model is around 9 to 1. The model utilizes 2048 tokens of a context window and full attention blocks. Currently, this is the most powerful model for the Russian language. The architecture of the model is a transformer decoder with 24 blocks, 16 heads each. The model is trained by predicting the next word in a sequence. Since we are working with a seq2seq problem, to generate output (a question), we need to feed the model with a proper prompt. Here is an example of a prompt and a response of the model:

(rationale)

[question word, key word]

-question

-answer

(rationale)

[question word, key word] (затравка) (NLTK – специализированная среда для автоматической обработки текстов, созданная для работы с Python)

[Что, NLTK]

-Что такое NLTK?

-Специализированная среда для автоматической обработки текстов, созданная для работы с Python. (В состав NLTK входит коллекция корпусов и словарные базы данных.)

[Что, состав]

(generation)

-Что включает в себя NLTK?

-NLTK включает в себя: -PyLib (Публичная библиотека Python, содержащая библиотеку, написанную на языке Python для обработки PHP ... (конец генерации, предел длины последовательности)

(rationale)
[questionword, keyword]
-question
-answer
(rationale)
[questionword, keyword]
(seed) (NLTK is a specialized environment for automated text processing with Python)
[What, NLTK]
-What is NLTK?

-a specialized environment for automated text processing with Python.
(NLTK contains collections of text corpora and dictionaries.)
[What, contains]
(generation)
-What are the parts of NLTK?
-NLTK contains PyLib (a public Python library that contains a module for PHP processing ... (finish generation, exceeded length limit)

2.3. Answer Evaluation

One common simple approach to the problem of answer evaluation relies on comparing word or lemmas symbol-by-symbol. We expand on this idea and attempt to search for semantically similar segments. This approach enhances the fuzzy string search, similar to how predicate comparison enhances the method for enumerating intersections of sets of semantic relationships.

Here is an explanation of the steps required to compare the answer given by a student to the correct answer. We assume that text fragments are represented by dependency trees; similar to depth-first search, both dependency trees are traversed: missing an edge results in a penalty, and the cost is an adjustable value based on several parameters, such as the length of text fragments and the complexity of the dependency trees; also, the cost may differ for the student's answer and the correct answer.

At each step, the nodes are compared by calculating the dot product between vector embeddings for the given words and scaling with an additional adjustable parameter; for example, the distance between the node and the root of the tree may imply a relative significance of a particular sub-tree.

Between all paths from the root to a node, the one with the highest cumulative similarity is selected. Those values can then be used to fine-tune the threshold for making a decision.

There are various ways to construct dependency trees; for our purposes, we chose to work with objects that hold attributes and pointers to other objects, such as word embeddings and links from parent nodes to children nodes. Additional attributes may include, for example, tags for parts of speech.

For experiments, we presented several teachers with a text fragment from learning materials about linguistics and asked them to formulate questions, and the teachers came up with a total of 26 questions. Then, we asked four students to answer the questions, and finally, we asked the teachers to arrange the answers from the most relevant to the least relevant (adding the correct answer forefront). During testing, we investigated the influence of the following parameters and the performance of the algorithm:

1. The scaling parameter for the distance between nodes;
2. The penalty for missing an edge in the dependency tree of the correct answer;
3. The penalty for missing an edge in the dependency tree of the student's answer;
4. The normalization coefficient for decision-making.

For this particular dataset, we acquired the following values:

1. The scaling parameter w_1 for the distance between nodes is calculated as

$$w_1 = \frac{1}{depth}, \qquad (2)$$

where *depth* is the distance between the root and a node

2. The penalty w_2 for missing and edge in the dependency tree of the correct answer is 2
3. The penalty w_3 for missing and edge in the dependency tree of the students answer is 0.6
4. The normalization coefficient w_4 for decision-making is calculated as

$$w_4 = \frac{1}{1 + \alpha(len_1 + len_2)}, \qquad (3)$$

where $\alpha = 0.05$, len_1 and len_2 are lengths of the correct and the students' answers, respectively.

Examples of answer evaluations are displayed in Table 7.

Table 7. Evaluations of answers (*distance* from the correct answer).

	Answer 1 (Correct)	Answer 2	Answer 3	Answer 4	Answer 5
Question 1	0.0	0.3	0.3	0.6	0.9
Question 2	0.0	0.5	0.6	0.9	0.9
Question 3	0.0	0.6	0.6	0.7	0.5
Average	0.0	0.7	0.8	1.0	1.4

It is important to note that, even with fine-tuning, for this dataset, we cannot establish thresholds that would allow separating segments for each answer in order (the segment for the first answer, the segment for the second answer, etc.). There might be various reasons for that, however, reviewing them here would be beside the point, since it was not the goal in the first place. For this particular problem, we look for an increase in value with a decrease in relevance while establishing a threshold (or lack thereof), which is of methodological interest.

2.4. System Architecture

To conform to the requirements of the specific problem we investigated, a virtual dialogue assistant for remote exams has to be considered in the context of design and technical implementation for online education platforms. The essential part is to provide clear external interfaces encapsulating internal processes and ensure the internal processes are manageable, maintainable, and scalable. To achieve that, our solution is to follow modern design principles for building information systems, separating the system into distributable modules implementing business-logic subroutines and front-end modules, providing application programming interfaces (API) for requesting specifications of datatypes and scenarios supported by the system and queuing tasks, such as generating question, evaluating answers, and so forth. One notable advantage of this approach is that the system can be fine-tuned to use-cases supported by a given online education platform. For any modern information system, its distributed nature (also implying scalability) is not as much a feature, but a necessity; therefore, considering its architecture, it is vital to make sure it can provide those. One of the key measurable characteristics of a system to evaluate its scalability is the complexity of interfaces: internal and external. In our description of the components of the system, we implicitly demonstrate that the components exhibit a high level of decoupling, proving confidence in that even with the increased sophistication of particular modules, it ought not to produce a necessity for a significant increase in complexity of interfaces.

3. Discussion and Conclusions

The goal of this research was to investigate solutions for enhancing the process of evaluation of students' performance at online education platforms. We purposely do not imply our interest in developing a general question-and-answer system, as such a proposition would necessarily require a much deeper and broader analysis of areas beyond the scope of this work. While restrictions, by definition, limit freedom, they also provide guarantees, allowing for a more focused view on a problem. Here, we limited ourselves to looking into traditional and modern approaches to find whether and in what manner they may be applicable for building a system for evaluation of students' performance, and we restricted the environment in which this evaluation is ought to be performed—specifically, online education platforms. With the environment specified, we now review various approaches and techniques for both solving specific sub-problems and organizing them together into a solution offering new opportunities or improving the existing ones. To our conclusion, it appears that both traditional techniques, requiring affordable computational resources and manual labor, and modern state-of-the-art methods relying on significantly large computational and storage resources, allow for building a system that can significantly improve performance evaluation at online education platforms. We believe that further improvements are primarily to be either of a quantitative nature (storage, computation, data mining, etc.) or of a methodological one.

Author Contributions: Conceptualization, A.M., O.M. and Y.M.; methodology, A.M., O.M. and Y.M.; software, A.M., A.S., P.K., A.R. and A.A.; validation, A.M., O.M., A.S., P.K., A.R. and A.A.; formal analysis, A.M., O.M. and Y.M.; investigation, A.M., O.M., A.S., P.K., A.R. and A.A.; resources, O.M. and Y.M.; data curation, O.M.; writing—original draft preparation, A.M. and O.M.; writing—review and editing, Y.M.; visualization, A.M. and O.M.; supervision, A.M., O.M. and Y.M.; project administration, A.M. and O.M.; funding acquisition, A.M. and O.M. All authors have read and agreed to the published version of the manuscript.

Funding: This research was financially supported by ITMO University, 197101 Saint Petersburg, Russia.

Conflicts of Interest: The authors declare no conflict of interest. The funders had no role in the design of the study; in the collection, analyses, or interpretation of data; in the writing of the manuscript, or in the decision to publish the results.

References

1. Mclsaac, M.; Gunawardena, C. Distance education. In *Handbook of Research for Educational Communication and Technology: A Project of the Association for Educational Communication and Technology*; Routledge: London, UK, 1996; pp. 403–437.
2. Rumble, G. Re-inventing distance education, 1971–2001. *Int. J. Lifelong Educ.* **2001**, *20*, 31–43.
3. Global Monitoring of School Closures. Available online: https://en.unesco.org/covid19/educationresponse (accessed on 14 May 2021).
4. Witze, A. Universities will never be the same after the coronavirus crisis. *Nature* **2020**, *582*, 162–164. [CrossRef]
5. E-Learning Global Market Trajectory & Analytics. Available online: https://www.strategyr.com/market-report-e-learning-forecasts-global-industry-analysts-inc.asp (accessed on 14 May 2021).
6. Digital Education Market by Learning Type (Self-paced and Instructor-led Online Education), Course Type, End-user (Individual Learners and Academic Institutions, Enterprise and Government Organizations), and Geography (North America, Europe, APAC and RoW)-Forecast up to 2026. Available online: https://reports.valuates.com/market-reports/INFO-Othe-4I48/digital-education (accessed on 20 May 2021).
7. Raupach, T.; Brown, J.; Anders, S.; Hasenfuss, G.; Harendza, S. Summative assessments are more powerful drivers of student learning than resource intensive teaching formats. *BMC Med.* **2013**, *11*, 31–43. [CrossRef]
8. Kirkpatrick, R.; Zang, Y. The Negative Influences of Exam-Oriented Education on Chinese High School Students: Backwash from Classroom to Child. *Lang. Test. Asia* **2011**, *1*, 36. [CrossRef]
9. Newble, D. Revisiting 'The effect of assessments and examinations on the learning of medical students'. *Med Educ.* **2016**, *50*, 498–501. [CrossRef] [PubMed]
10. Perera-Diltz, D.M.; Moe, J.L. Formative and Summative Assessment in Online Education. *J. Res. Innov. Teach.* **2014**, *7*, 130–142.
11. Chaudhary, S.V.S.; Dey, N. Assessment in Open and Distance Learning System (ODL): A Challenge. *Open Prax.* **2013**, *5*, 207–216. [CrossRef]
12. Halova, E.Y.; Kobilarov, R.G. Advantages and Disadvantages of the Test Method for Checking and Evaluating of the Knowledge, the Skills and the Habits of Students. *AIP Conf. Proc.* **2010**, *1203*, 1325–1328. [CrossRef]
13. Indurkhya, N.; Damerau, F.J. *Handbook of Natural Language Processing*, 2nd ed.; CRC Press: Boca Raton, FL, USA, 2010.
14. Clark, A.; Fox, C.; Lappin, S. *The Handbook of Computational Linguistics and Natural Language Processing*; John Wiley & Sons: Hoboken, NJ, USA, 2010; doi:10.1002/9781444324044. [CrossRef]
15. Otter, D.W.; Medina, J.R.; Kalita, J.K. A Survey of the Usages of Deep Learning in Natural Language Processing. *arXiv* **2019**, arXiv:cs.CL/1807.10854.
16. An overview of empirical natural language processing. *AI Mag.* **1997**, *18*, 13.
17. Hutchins, W.J. Machine Translation: A Brief History. In *Concise History of the Language Sciences*; Elsevier: Amsterdam, The Netherlands, 1995. [CrossRef]
18. Estival, D.; Nowak, C.; Zschorn, A. Towards ontology-based natural language processing. In Proceeedings of the Workshop on NLP and XML (NLPXML-2004): RDF/RDFS and OWL in Language Technology, Barcelona, Spain, 25 July 2004. [CrossRef]
19. Huang, Y.; Cheng, Y.; Bapna, A.; Firat, O.; Chen, M.X.; Chen, D.; Lee, H.J.; Ngiam, J.; Le, Q.V.; Wu, Y.; Chen, Z. GPipe: Efficient training of giant neural networks using pipeline parallelism. *Adv. Neural Inf. Process. Syst.* **2019**, *32*, 103–112.
20. Alexeev, A.; Kukharev, G.; Matveev, Y.; Matveev, A. A Highly Efficient Neural Network Solution for Automated Detection of Pointer Meters with Different Analog Scales Operating in Different Conditions. *Mathematics* **2020**, *8*. [CrossRef]

21. Torfi, A.; Shirvani, R.A.; Keneshloo, Y.; Tavaf, N.; Fox, E.A. Natural Language Processing Advancements By Deep Learning: A Survey. *arXiv* **2021**, arXiv:cs.CL/2003.01200.
22. Keneshloo, Y.; Shi, T.; Ramakrishnan, N.; Reddy, C.K. Deep Reinforcement Learning for Sequence-to-Sequence Models. *IEEE Trans. Neural Netw. Learn. Syst.* **2020**, *31*. [CrossRef]
23. Hazelwood, K.; Bird, S.; Brooks, D.; Chintala, S.; Diril, U.; Dzhulgakov, D.; Fawzy, M.; Jia, B.; Jia, Y.; Kalro, A.; et al. Applied Machine Learning at Facebook: A Datacenter Infrastructure Perspective. In Proceedings of the 2018 IEEE International Symposium on High Performance Computer Architecture (HPCA), Vienna, Austria, 24–28 February 2018. [CrossRef]
24. Wang, F.Y.; Zhang, J.J.; Zheng, X.; Wang, X.; Yuan, Y.; Dai, X.; Zhang, J.; Yang, L. Where does AlphaGo go: From church-turing thesis to AlphaGo thesis and beyond. *IEEE/CAA J. Autom. Sin.* **2016**, *3*. [CrossRef]
25. Guarino, N. *Formal Ontology in Information Systems: Proceedings of the 1st International Conference, Trento, Italy, 6–8 June 1998*; IOS Press: Trento, Italy, 1998; Volume 46.
26. Guizzardi, G. Ontology, Ontologies and the "I" of FAIR. *Data Intell.* **2020**, *2*, 181–191. [CrossRef]
27. Wolfram Data Framework. Available online: https://www.wolfram.com/data-framework (accessed on 15 July 2021).
28. how many goats in spain. Available online: https://www.wolframalpha.com/input/?i=how+many+goats+in+spain (accessed on 15 July 2021).
29. Das, B.; Majumder, M.; Phadikar, S.; Sekh, A. Automatic generation of fill-in-the-blank question with corpus-based distractors for e-assessment to enhance learning. *Comput. Appl. Eng. Educ.* **2019**, *27*, 1485–1495. [CrossRef]
30. Rakić, K. The proposal of the intelligent system for generating objective test questions in controlled natural language for domain knowledge based on ontology. In Proceedings of the 2016 International Conference on Smart Systems and Technologies (SST), Osijek, Croatia, 12–14 October 2016; pp. 135–138. [CrossRef]
31. Marrese-Taylor, E.; Nakajima, A.; Matsuo, Y.; Yuichi, O. Learning to Automatically Generate Fill-In-The-Blank Quizzes. In Proceedings of the 5th Workshop on Natural Language Processing Techniques for Educational Applications; Association for Computational Linguistics, Melbourne, Australia, 19 July 2018; pp. 152–156. [CrossRef]
32. CH, D.R.; Saha, S.K. Automatic Multiple Choice Question Generation From Text: A Survey. *IEEE Trans. Learn. Technol.* **2020**, *13*, 14–25. [CrossRef]
33. Kurtasov, A. A System for Generating Cloze Test Items from Russian-Language Text. In Proceedings of the Student Research Workshop Associated with RANLP 2013, Hissar, Bulgaria, 9–11 September 2013; INCOMA Ltd. Shoumen, BULGARIA: Hissar, Bulgaria, 2013; pp. 107–112.
34. Wang, D.; Zhao, Y.; Lin, H.; Zuo, X. Automatic scoring of Chinese fill-in-the-blank questions based on improved P-means. *J. Intell. Fuzzy Syst.* **2021**, *40*, 5473–5482. [CrossRef]
35. Batura, T.; Charintseva, M. *Basics of Text Information Processing*; A.P. Ershov Institute of Informatics Systems (IIS) SB RAS: Novosibirsk, Russia, 2016.
36. Solovyev, A. Syntactic and Semantic Models and Algorithms in Question Answering. In Proceedings of the 13th All-Russian Scientific Conference "Digital libraries: Advanced Methods and Technologies, Digital Collections", RCDL 2011, Voronezh, Russia, 19–22 October 2011.
37. Schlaefer, N. *A Semantic Approach to Question Answering*; Verlag Dr. Müller: Riga, Latvia, 2011.
38. Solovyev, A. Dependency-based algorithms for answer validation task in Russian question answering. In *Language Processing and Knowledge in the Web*; Springer: Berlin/Heidelberg, Germany, 2013; Volume 8105. [CrossRef]
39. Scharpf, P.; Schubotz, M.; Youssef, A.; Hamborg, F.; Meuschke, N.; Gipp, B. Classification and clustering of arxiv documents, sections, and abstracts, comparing encodings of natural and mathematical language. In Proceedings of the ACM/IEEE Joint Conference on Digital Libraries in 2020, Virtual Event, China, 1–5 August 2020. [CrossRef]
40. Krstovski, K.; Blei, D.M. Equation Embeddings. *arXiv* **2018**, arXiv:stat.ML/1803.09123.
41. Bogoradnikova, D.; Makhnytkina, O.; Matveev, A.; Zakharova, A.; Akulov, A. Multilingual Sentiment Analysis and Toxicity Detection for Text Messages in Russian. In Proceedings of the 2021 29th Conference of Open Innovations Association (FRUCT), Tampere, Finland, 12–14 May 2021. [CrossRef]
42. Laptev, A.; Andrusenko, A.; Podluzhny, I.; Mitrofanov, A.; Medennikov, I.; Matveev, Y. Dynamic acoustic unit augmentation with bpe-dropout for low-resource end-to-end speech recognition. *Sensors* **2021**, *21*, 3063. [CrossRef] [PubMed]
43. Bruches, E.; Pauls, A.; Batura, T.; Isachenko, V. Entity Recognition and Relation Extraction from Scientific and Technical Texts in Russian. In Proceedings of the 2020 Science and Artificial Intelligence Conference (SAI Ence), Novosibirsk, Russia, 14–15 November 2020. [CrossRef]

Article

Ubiquitous Computing: Driving in the Intelligent Environment

Emanuela Bran [1,2,*], Elena Bautu [3], Dragos Florin Sburlan [3], Crenguta Madalina Puchianu [3] and Dorin Mircea Popovici [1,3]

[1] Faculty of Electrical Engineering and Computer Science, "Transilvania" University, 500024 Brasov, Romania; dmpopovici@univ-ovidius.ro
[2] Institute of Studies for Development and Security at the Black Sea, "Ovidius" University, 900527 Constanta, Romania
[3] Department of Mathematics and Computer Science, "Ovidius" University, 900527 Constanta, Romania; ebautu@univ-ovidius.ro (E.B.); dsburlan@univ-ovidius.ro (D.F.S.); crenguta.puchianu@univ-ovidius.ro (C.M.P.)
* Correspondence: emanuela.bran@365.univ-ovidius.ro; Tel.: +40-7240-21149

Citation: Bran, E.; Bautu, E.; Sburlan, D.F.; Puchianu, C.M.; Popovici, D.M. Ubiquitous Computing: Driving in the Intelligent Environment. *Mathematics* 2021, 9, 2649. https://doi.org/10.3390/math9212649

Academic Editors: Grigoreta-Sofia Cojocar, Adriana-Mihaela Guran and Bo-Hao Chen

Received: 31 July 2021
Accepted: 15 October 2021
Published: 20 October 2021

Publisher's Note: MDPI stays neutral with regard to jurisdictional claims in published maps and institutional affiliations.

Copyright: © 2021 by the authors. Licensee MDPI, Basel, Switzerland. This article is an open access article distributed under the terms and conditions of the Creative Commons Attribution (CC BY) license (https://creativecommons.org/licenses/by/4.0/).

Abstract: In the context of hyper-connected cars and a growing heterogeneous digital ecosystem, we wish to make the most of the data available from the various sensors, devices and services that compose the ecosystem, in order to propose a proof of concept in-vehicle system that enhances the driving experience. We focus on improving the driving experience along three main directions, namely: (1) driving and trip planning, (2) health and well-being and (3) social and online activities. We approached the in-vehicle space as a smart interface to the intelligent driving environment. The digital data-producers in the ecosystem of the connected car are sources of raw data of various categories, such as data from the outside world, gathered from sensors or online services, data from the car itself and data from the driver gathered with various mobile and wearable devices, by means of observing his state and by means of his social media and online activity. Data is later processed into three information categories—driving, wellness, and social—and used to provide multi-modal interaction, namely visual, audio and gesture. The system is implemented to act in response to the trafficked information on different levels of autonomy, either in a reactive manner, by simple monitoring, or in a proactive manner. The system is designed to provide an in-vehicle system that assists the driver with planning the travel (*Drive panel*), by providing a comfortable environment for the driver while monitoring him (*Wellness panel*), and by adaptively managing interactions with their phone and the digital environment (*Social panel*). Heuristic evaluation of the system is performed, with respect to guidelines formulated for automated vehicles, and a SWOT analysis of the system is also presented in the paper.

Keywords: ubiquitous computing; smart cars; natural interfaces; multimodal interaction; smart devices; gesture input; voice input

1. Introduction

Complex in-vehicle software systems are a hallmark of premium cars, augmenting the driver's experience at many different levels. In today's interconnected world, connected cars should offer safer trips and a more pleasant journey for the driver and the passengers altogether. Intelligent cars, wearable devices, mobile devices, smart cities and the digital environment will all be connected in order to improve the driving user experience.

The smart cars [1] sector is increasing tremendously as a part of the internet of things (IoT). Connected cars provide intelligent advanced driver assistance systems (ADAS), and 5G communication with smart road infrastructure [2]. Data-driven business models [3] are created for intelligent transportation systems and beyond. The in-vehicle space is populated with smart devices such as wearables and insideables, that are smart sensors worn by people or implanted inside the human body [4]. These devices serve as natural interfaces for communication with the in-vehicle systems (IVIS) [5].

Other smart devices and services are integrated and connected to smart cars in order to create systems of systems (SoS) that serve complex driving related tasks [6]. These networks of systems give rise to smart cities with intelligent mobility services [7] that have an important impact on sustainability [8]. The intelligent environment is interconnected from production, to transportation, to the human sector and beyond, in a planetary space where the physical and the digital intertwine creating global intelligence [9].

We may view the myriad of the actors and stakeholders of the digital environment as part of a greater ecosystem with interdependent links [10]. These IoT innovations are possible with the help of 5G technology over which speed dependent applications flourish, and homes, cities cars become smart entities creating ambient intelligence [11]. All these breakthrough technologies emerging inside smart vehicles create an intuitive and extraordinary driving experience of the digital transformation sage [12].

We aim to design a system that relies on design thinking principles as we consider a user-centered design, and we encompass the driver's emotional, cognitive, and aesthetic needs while he is driving and handling an intelligent environment. The goal of the paper is to propose a proof-of-concept system that is built on top of various open source or free libraries/APIs/SDKs (described in Sections 3.2 and 3.3) and incorporates multimodal interaction (voice, gesture, touch, pulse, facial emotion etc.) to enhance the driving experience. We consider the interior of the car as both a multi-modal interface able to process various user input, and a part of the intelligent environment. In this respect, having in mind the design of a useful, supportive, and comfortable environment for the driver and passengers (the "digital car-sphere"), we propose a paradigm-shift away from conventional/standard in-vehicle user interfaces. More precisely, in the proposed design, the entire in-vehicle digital space sphere represents an interface for the intelligent digital medium. Various onboard or outboard sensors, devices, and systems gather, integrate, and process data, providing in return relevant information and proactive interaction with the requesters (the driver, passengers or other systems). The human actions on various time scales (ranging from simple instant tactile interactions with the standard input devices to short/long time behavior or even voice) are transduced into data which can be interpreted by the intelligent environment in order to adaptively trigger a better in-vehicle experience for the requester. In a typical scenario, as the driver/passenger engages the in-vehicle activities, the intelligent in-vehicle medium will track its actions in order to support intelligent adaptation to its ongoing task(s). In our setup this is done by considering four main components: data, information, interaction, and autonomy. In our view the data refers to values collected together from the entire driving environment (in- or out-vehicle). These values are classified by the information component into several categories (driving, wellness, social). Based on this classification the multimodal interface incorporates different types of interactions tailored to the user's perceptual senses. Finally, autonomy refers to the digital environment feed-back as a consequence to driver/passenger actions and other in- or -out vehicle data collection.

In this article we focused on the in-vehicle space as an interface [13] to the intelligent environment. Our vision is that the in-vehicle environment should offer assistance [14,15] for the user in three main areas, namely: (1) the driving and trip planning task, (2) their health and well-being and (3) managing social connections and online activities. By empowering the driver to have access to information ecosystem [16] regarding these three areas and providing multi-modal communication and implementing proactive interaction for the in-vehicle system, we aim to create an enhanced driving experience that also diminishes the overloading of the driver with information unrelated to the driving task.

The paper is structured in five sections. Section 2 covers related works in the field of in vehicle systems and ubiquitous computing. The brief review of the current state of the art in the field is organized in three subsections, each dealing with a different focus, as follows: the first subsection reviews targeted literature on connected cars and smart transport infrastructure, the second subsection reviews targeted literature on wearable devices and the intelligent world environment, and finally, the third subsection of the

Literature review deals with papers on the topic of user experience and car multimodal interfaces. Section 3 presents the proposed system, detailing the three main directions in which it touches upon the driving experience. The multi-modal interaction components are described in this context. The results of heuristic evaluation of the systems according to human machine interface (HMI) guidelines for automated vehicles are discussed in Section 4. Finally, the paper ends with our conclusions and outlines some avenues for future research.

2. Literature Review

This paper presents a proof-of-concept in-vehicle system placed in the ubiquitous computing context. To this end, the literature review that follows covers aspects that relate to connected cars and smart infrastructure, wearables and the intelligent environment, user experience and car multimodal interfaces. We structured the literature review in 3 main categories of studies that offer a broad perspective on in-vehicle systems in the interconnected and complex digitalized world. We will start by going through communication innovations of connected cars. From there we will investigate other connection endpoints around the intelligent environment. Finally, we will get an insight into natural interfaces. Altogether, we wish to understand and present the interior of the car as both an interface and a part of the intelligent environment.

2.1. Connected Cars and Smart Transport Infrastructure

Whether we are considering autonomous vehicles [17], or cars with a certain degree of automation in general, there is always a necessity for the car to be connected to an information provider of its surroundings, in order to either inform the intelligent car system or the human driver. Cars may be connected [18] in a variety of ways, a simple navigation system [19] being one of the first examples of connectivity. Today, there are cars that connect to the internet using a SIM card, and applications [20] that require fast communication are being designed to use the benefits of the 5G [21,22] and IoT technology [23,24], creating an internet of vehicles digital space [25,26]. There are also other ways and protocols [27] that cars use to communicate between them, such as through blockchain secured ad-hoc vehicle networks [28,29] and LiFi communication [30,31].

Smart cars [32] connect to each other through what is called V2V communication [33] to synchronize traffic and safely [34] carry out driving actions. They may also connect to the road infrastructure in V2I communication [35,36] and generally connect to other entities in V2X communication [37,38] such as to interact with pedestrians [39]. Smart cities are already implementing such communications that take place between different smart entities, some of which are traffic related. There is research going on to enable connected cars to coordinate with each other in a decentralized proactive manner as opposed to just request information [40]. Connected car applications [41] already in use provide means for the car to connect to an edge-fog-cloud of information [42–44] for various purposes in a centralized manner.

Based on research on autonomous or non-autonomous cars, on different communication technologies, on decentralized and centralized connectivity [45,46], we have concluded that there is a rich cluster of information and interaction produced between vehicles and other road entities. In the Solution section, we are going to explore a dynamic visualization [47] of road information in terms of predictive weather conditions [48], visibility [49,50] and outdoor illumination [51,52] and potential sun glares [53,54] at the future moment of passing through that area, with markings on possible hazards [55,56], for a better trip planning. We chose this kind of information compared to the better-known real-time crowd sourced [57] traffic data, as an example of a novel cluster of visualized road information.

2.2. Wearables and the Intelligent World Environment

Smart sensors [58] are used everywhere, ranging from smart devices to the human body and the environment. There is even the term "EveryWare" referring to ubiquitous computing, coined by Greenfield, (Greenfield A (2006) Everyware: The dawning age of ubiquitous computing. New Riders, Berkeley) [59,60]. Wearables [61,62] are devices that can be worn by people and have smart sensors incorporated that collect information on the vital signs [63], the whereabouts, the motion [64], and other environmental conditions. They may come in a more traditional form such as a wristband, an innovative form such as a nail sticker [65], a tattoo [66], or even a smart textile [67–70], and they contribute to the health [71] and wellbeing [72,73] of people.

Smart sensors can be very small; they are able to communicate and may process data. Together with smart actuators [74,75] they may sense the environment, modify the environment, and synchronize with each other [76,77]. They make ubiquitous computing possible, which is a paradigm under which services follow the user seamlessly across different environments, comprising altogether the intelligent environment. Computers started as one big static device used by many users, then they became smaller and affordable in the form of personal computers, then one user could own many mobile computer devices. Now the idea is to extend the number of devices of the user by the thousands, embedding them across the environment [78], making them sharable as hardware, and personalized as software [79].

In the Proposed Solution section, we will explore the potential of a wearable pulse sensor [80] for monitoring the human body [81], and the OBDII interface to access data from the car's sensors. For creating a pleasant ambience to suit the driver's needs, we will also have playlists of songs [82] that have been analyzed in terms of beat rate, liveliness, and genre as an example of data request. In addition, the weather data that we use is also the product of a multitude of smart things (satellites, ground sensors), collecting big data, collaboratively processing it, and finally delivering it through an interface designed for people or machines in the form of an API.

2.3. User Experience and Car Multimodal Interfaces

Intelligent environments may use mixed reality for visualization [83,84] and natural interaction metaphors [85] that let the user intuitively use the system. They identify and monitor the user [86], read intentions [87], and personalize their services. By also monitoring the environment [88], they infer circumstances, and adapt [89] their services accordingly. They act as autonomous and affective agents [90–92], understanding the surroundings [93], initiating interaction, and completing tasks on their own. Intelligent environments are proactive in contrast with other reactive systems creating a state-of-the-art user experience [94–97] by emulating humans [98] and fulfilling desires [99].

The car, as part of the intelligent environment [100], should function as an empathic agent [101], having a multimodal interface [102] making use of artificial intelligence [103], mixed reality [104], and natural interaction [105,106]. Multimodality [107] uses different channels and modes of communication. Through the visual field we may understand text, symbols, images, animations, and the system may capture gestures, recognize objects, and detect the depth of the environment. Sound and touch represent other widely used channels of communication, and the list may continue, especially for the unlimited possibility of digital systems to augment their perception through a wide variety of sensors. As for humans, there is research going on into EMG, EOG, EEG interfaces [108]. Brain interfaces [109,110] may even infer the state of the mind by analyzing brain waves and augment expression of intensions.

We proposed and designed several types of multimodal natural interaction in the Solution section, between the intelligent in-vehicle system and the driver. The system recognizes facial expressions [111], speech, gestures, and touch. As it is believed to be a correlation [112,113] between music and the emotional state [114] or the physiological alertness [115], a specific playlist is suggested [116,117]. Notifications from the smart phone are also synchronized with the in-vehicle system to better manage driver-phone interactions [118]. By being aware of certain stressful [119] circumstances the system can decide whether to postpone notifications in an attempt to increase safety [120].

3. Proposed Solution

This section details a presentation of the system, with emphasis on some conceptual, technical and implementation details. Our aim is to help the driver concentrate on the driving task by planning the travel (*Drive panel*, see Section 3.2.1), by providing a comfortable environment for the driver while monitoring him (*Wellness panel*, see Section 3.2.2), and by managing interactions with their phone and the digital environment (*Social panel*, see Section 3.2.3).

Ubiquitous Computing is a paradigm under which services follow the user seamlessly across the intelligent environment [121,122]. We apply this paradigm by considering that the driver is moving through the intelligent environment and services are continuously adapting to the present circumstances. The intelligent driving environment (detailed in Section 3.1) is comprised by the in-vehicle space, the nearby surroundings, and faraway elements that are remotely connected, all of which have a direct or an indirect impact on the here and now.

The in-vehicle space acts as an interface for the intelligent environment. It harvests and hosts data, which is aggregated and processed into information, building an information ecosystem. This interface enables the user to naturally interact with data and information by providing several types of multimodal interaction such as touch, voice, and gesture. As an intelligent interface, the system was designed to be proactive. Thus, it decides the moment and means to notify the user and initiates interaction when necessary. It achieves this by analyzing the priority of the information to be communicated and by inferring the status of the user.

The smart driving interface components (see Figure 1) are data (further described in Section 3.1), information (further described in Section 3.2), interaction (further described in Section 3.3), and autonomy (further described in Section 3.4). We would like to emphasize the fact that the list of data presented in this article is not exhaustive, but it is merely a fraction of the types of data available, as discussed in the State-of-the-Art section. By choosing our data producers, we meant to exemplify data categories (world, car, driver), that are later processed into information categories (driving, wellness, social), how they help provide multimodal interaction (visual, audio, gesture), and how they help the system act on different levels of autonomy (reactive, monitoring, proactive).

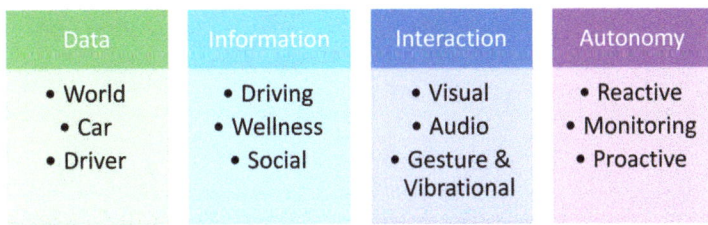

Figure 1. The smart driving interface components.

3.1. Harvested Categories of Data

From our perspective, the driving environment is very abundant in data, either requested or locally produced, with a broad origin spectrum, which effectively processed by correlation and aggregation, create a vast information field that is rich in meaning. In this subchapter we are going to analyze the devices and services from where we harvest data that is relevant to the driving intelligent environment. We will organize the data sources by origin into layers of different categories and subcategories.

In Figure 2, the intelligent driving environment layers are presented through a user centered design. The data producers are grouped by their data source into four layers, easily distinguishable in the figure by their color (orange for the user signals, green for the interceding IVIS, purple for the smart devices, blue for the internet services). The person sits at the center of the diagram, surrounded by the awareness of the IVIS which intercedes between him and the environment, that is comprised of the nearby car interior populated by smart devices and the faraway services present in the internet cloud.

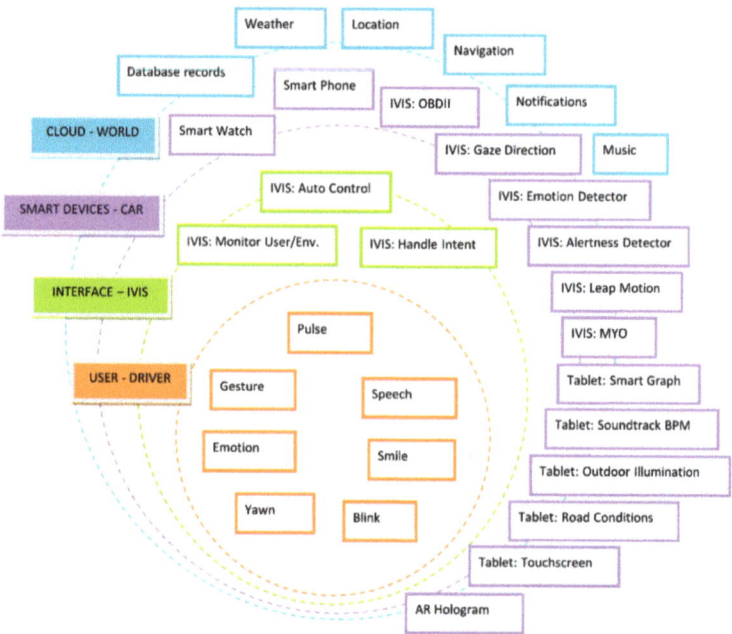

Figure 2. Intelligent driving environment layers.

3.1.1. The Internet Cloud (the World)

We harvest data from the cloud to provide information on the weather and environmental conditions that have an impact on the difficulty of driving. Data about navigation routes also falls under this category. The music that plays in the car is also broadcasted from the internet, while notifications usually are also related to events happening in the real or digital world. We should also mention here the database records that are stored in the cloud. The location of the mobile phone is also dependent on mobile services.

3.1.2. The Smart Devices (the Car)

The smart devices inside the car are producers of data, and we can list here the OBDII (http://www.obdii.com/, accessed on 1 July 2021) interface for vehicle diagnosis and reporting, which reads the values indicated by the speed, rotations per minute, oil, and water sensors. Some smart devices are linked to the people inside the cars, and these are the Smart Watch (which reads the pulse, provides voice input interaction, and vibrational feedback) and the Smart Phone (which send notifications, and provides location information).

Other devices from inside the car that produce data are the Leap Motion device (which reads gestures), the camera (that sends video footage for face detection). These two, together with the OBDII device, are linked to an onboard computer laptop which hosts the NodeJS server [123,124] of the system (which handles communication between data producers and consumers, together with database connection services). It handles emotion and alertness state detection, gaze direction calculation, gesture interpretation, and car sensor data collection.

The visual interface is provided by a touch-enabled smart tablet display. It also handles the calculation of the Sun's position and integrates this with navigation route data and weather data to obtain road conditions. It listens to data coming from sensors and queries data from the database in order to plot graphs. It is the place where multimodal input is centralized for interaction with the displayed elements, and where automatic control is carried out by centralizing user and environment monitoring data. We have also experimented with an AR hologram floating right past the windshield [125]. This was achieved by using a video projector and a screen which was reflected by the windshield.

3.1.3. The Ubiquitous Interface (the IVIS)

The intelligent interface is provided by the IVIS which processes raw data coming from the world, the car, and the user. This layer handles interaction between the user, the car's interior environment and the outside real and digital world, by providing detection of the user and the environment, reaction of the system to the user's intentions, and proactive automated action towards the inferred user's needs and the environment's circumstances. It is composed by the onboard computer which handles resource costly calculations, and the tablet which centralizes information. Interaction is further discussed in Section 3.3 and autonomy in Section 3.4.

3.1.4. The Monitored User (the Driver)

The user is a provider of data regarding physiological parameters and communication through multimodal interaction. The facial expressions indicate emotional status and alertness level, and the system watches for blinking patterns, yawns, emotions, and smile through the camera. We have also experimented with the gaze direction of the user as an indication of where their focus of attention is located [126]. The user's gestures are captured by the Leap Motion and MYO devices, and their voice and pulse by the Smart Watch.

3.2. Processed Categories of Information

We have identified so far, a vast spectrum of data that is relevant to the in-vehicle intelligent environment. We will now proceed by explaining how data is aggregated into clusters linked by meaning and purpose. By further correlating and processing data, we obtain several categories of information. In this subchapter we are going to present these categories and how they are obtained.

3.2.1. Car, Road and Driving Task Related Information

In Figure 3 we present the *Driving task panel* for when we start a trip from the city of Constanta to Brasov city. On the left side there is a map augmented by weather and illumination information. On the right side, we output information coming from the OBDII interface, composed of either an icon, or a text extracted from a JSON with values for different parameters.

The route is obtained from the MapBox navigation and map graphics API service (https://docs.mapbox.com/api/overview/, accessed on 1 July 2021), which responds to a request having parameters such as route type and endpoints. The JSON response contains geolocation, time from departure and action indication of every turning point along the route. We processed this data into a set of segments calculating orientation for each of them. As input data consisted of geographical coordinates (latitude and longitude), we used special formulas (https://www.movable-type.co.uk/scripts/latlong.html, accessed on 1 July 2021) to determine the geodesic distance (shortest possible line between two points on a curved surface) and bearing (also called forward azimuth, here we used the formula for the initial heading angle).

The Haversine formula (1) is used for computing the length of a segment of the driving route, where φ is latitude, λ is longitude, R is Earth's radius (mean radius = 6371 km):

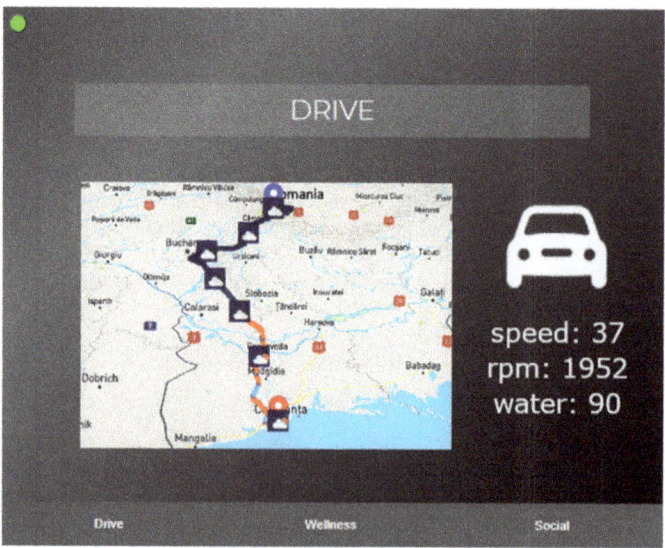

Figure 3. The *Drive panel*.

The term a is denoted by the formula:

$$a = \sin^2(\Delta\varphi/2) + \cos\varphi_1 \times \cos\varphi_2 \times \sin^2(\Delta\lambda/2)$$

and c by:

$$c = 2 \times a\tan(\sqrt{a}, \sqrt{(1-a)})$$

Then the distance d is computed as:

$$d = R \times c \tag{1}$$

The initial bearing, Equation (2), is used to compute the heading angle of a segment of the drivers' route φ_1, λ_1 is the start point, φ_2, λ_2 the end point ($\Delta\lambda$ is the difference in longitude):

$$\theta = a\tan2(\sin\Delta\lambda \times \cos\varphi_2, \cos\varphi_1 \times \sin\varphi_2 - \sin\varphi_1 \times \cos\varphi_2 \times \cos\Delta\lambda) \tag{2}$$

We used the JavaScript SunCalc library (https://github.com/mourner/suncalc, accessed on 1 July 2021) for calculating the sun altitude (angle of ground elevation) and azimuth (horizontal angle from north direction) for each location and time. This is a complex calculation process (https://www.aa.quae.nl/en/reken/zonpositie.html#9, accessed on 1 July 2021) which takes into account the geographical location and observation date and time, with all the planetary motions described by astronomy. By combining the obtained data, we inferred the natural illumination along the route, as well as certain road segments where there is a sun glare visibility hazard. We colored the route with different shades of blue according to the sky color at that time and location, and with shades of orange for sunrise and sunset glares coming through the windshield or through the mirrors.

We determined whether there is a risk of sunglare by using data from navigation and sun position the following way. If the Sun's altitude was below 30 degrees above the horizon, and if there was a smaller than 60 degrees difference of angles between the Sun's azimuth and the car's bearing, then there could be sunglare coming from the front. We computed in a similar way glares coming from the back of the car through the mirrors.

We also used the OpenWeather API service (https://openweathermap.org/api, accessed on 1 July 2021), which is based on machine learning for forecasting, to request weather conditions every 50 km along the route. The JSON response contains icons, and values for different parameters, for minutely, hourly, and daily forecast. We used this data to show the forecast along the route at the future time of passing through that location. These icons are spread across the route, and they alternate between time, temperature, weather conditions, and hazards (fog, high amount of precipitation, ice and snow, extreme temperatures, strong wind).

3.2.2. Personal Health and Wellbeing Related Information

The *Wellness-related panel*, shown in Figure 4, handles information connected to the wellbeing of the driver. The first icon represents the pulse, the second one the alertness inferred from facial recognition, and the last one the tempo of the music being played. Colors change in a fluid manner as to indicate the exact value of a parameter.

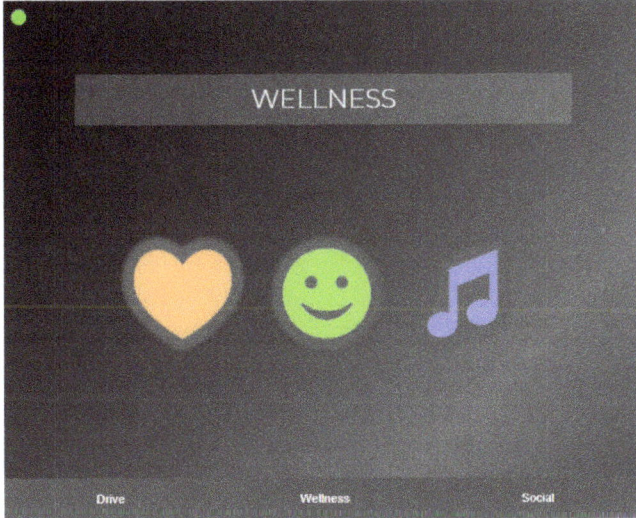

Figure 4. The *Wellness panel*.

Warm colors (*red* for maximum) express a high pulse, a nervous expression, and an energic playlist, while cool colors (*blue* for minimum) express a lower pulse, a tired expression (yawning or blinking at a high rate or too slow as time interval measured between closing and reopening the eye), and a slow playlist. A *green* pulse and a *green* emoji represent the optimal state. The pulse is constantly updated, while the tempo is updated once a new soundtrack is played.

Music is played with the help of the YouTube Player API (https://developers.google.com/youtube/iframe_api_reference, accessed on 1 July 2021), by uploading the code of each music track. The tempo is determined for each song using data from GetSongBPM API (https://getsongbpm.com/api, accessed on 1 July 2021), and a mean tempo is calculated for each musical track, considering that the songs will likely belong to the same genre. Then, music tracks are ordered by tempo and the one in the middle is the one played by default. Soundtracks with a greater tempo are intended to stimulate a driver and keep him alert, while slower soundtracks are intended to calm down a stressed driver.

The emoji is *green* by default and changes to *blue* when either a yawn is detected, or the eye blink rate indicates tiredness. It also changes to *red* when emotions are heightened such as in the case of anger, fear, or surprise. After a change, it slowly goes back to *green*. Facial feature detection and tracking is performed using the Beyond Reality Face SDK version 5 (https://github.com/Tastenkunst/brfv5-browser, accessed on 1 July 2021), which helped us detect yawn and patterns of blinking for tired eyes. The JavaScript FaceApi (https://justadudewhohacks.github.io/face-api.js/docs/index.html, accessed on 1 July 2021), which is built on the tensorflow.js core API, can discriminate between seven different facial expressions: neutral, happy, sad, angry, fearful, disgusted, and surprised. Only in the absence of tiredness (because a yawn might be interpreted as an angry face by the emotion recognition component), the facial expressions indicating negative emotions were counted as heightened levels of stress. This way the system clearly discriminates between tiredness, which is also a threat of higher priority, and angriness, which compared to tiredness is of a lower priority.

We have also designed graph plotting for the pulse (see Figure 5). The plotting function has several parameters. The first sets over what time interval we wish to show values. The seconds sets the time subunit for which we want to calculate a mean value and a variance value. If this parameter is set to 0 then the pulse is plotted for each numerical value and there is no variance. We experimented with different visualizations and called the function on different intervals such as, the pulse during last 3 h with segments representing the mean value and variance over 9 min subunits, and the pulse during the last 15 min with segments representing the mean and variance over 45 s subunits. Each segment end represents the mean value in a subunit interval. Variance coming from positive values is plotted above the graph while that produced by negative values is plotted under the graph. The numerical value is the pulse rate at the present moment.

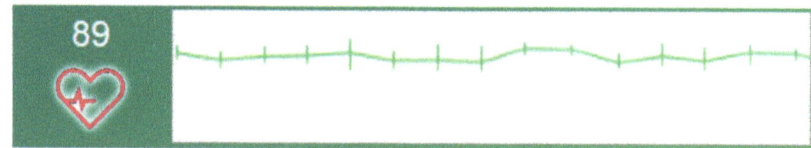

Figure 5. Plotting the mean pulse of the driver, while the vertical segments represent the variance.

3.2.3. Social and Media Updates Related Information

For the *Social information panel*, we have designed a double spiral menu for navigating notifications received from the smart phone. We chose this design because it helps display many items, they are ordered chronologically, nearby items are also close along the timeline, and it offers a fluid aesthetic.

The selected icon is centered and different text fields from the notification on the Android phone, value of attributes of the NotificationCompat object, are being shown on the right side, such as title, text, subtext, and application. We have assigned several well-known applications to different predefined categories, distinguished by color codes. Red represents the communication category, blue the social category, and violet other activities category.

In Figure 6, these categories are exemplified by Gmail, Facebook, and the Wish shopping app, respectively. Every time a new notification is received, it will be added to the list, and it will be automatically centered. Thus, notifications are chronologically ordered. Only 19 notifications are visible at a time, the last five on each spiral end fading into transparency. Notifications may be deleted after they are checked.

We have already experimented with interaction initiated from the tablet to the phone through the Euphoria server that is hosted on the laptop computer. We built two SOS buttons for local emergency services, one for calling the 112 line (112—the Romanian emergency phone number), and one for sending a SMS to the 113 line (designed for people with speech accessibility problems). When we clicked the call option, an Android Intent was produced on the phone that resulted in the android calling app being started and the 112 telephone number already typed. The user just needed to press call on their phone. When we clicked the SMS option, an Android Intent generated an SMS to the 113 (113—

the Romanian SMS emergency number) number with the text body containing personal information, a generic message asking for help, and the location information available from the mobile services. The user just needed to press send on their phone.

Figure 6. The *Social panel*.

Future work will include implementing the ability to initiate actions from the *Social panel* to the smart phone through the Android Intent system, such as making a call and responding to a WhatsApp message. The iOS also provides a similar capability based on App Extensions system.

3.3. Natural Multimodal Interaction

Multimodal interaction is a key feature in ubiquitous computing, and we provide three distinct channels of interaction between the user and the system. We implemented for each of these channels (visual, audio, gesture/vibrational) both input and output communication. Another key feature of ubiquitous computing is actively monitoring users. Thus, some of these modes of interaction are active while some are passive from the user's side, making the system reactive as well as proactive. Proactiveness will be further discussed in Section 3.4.

Each panel has an interactive element that may be changed through quick multimodal interaction (see Figure 7). For the *Driving panel*, this element is the departure time. When planning a trip, the user may want to consider how the conditions will be if he stops for a while along the route or leaves a later moment in time. By selecting the panel, the user may easily scroll the timeline of the map to make better decisions ahead. For the *Wellness panel*, the interactive element is the music, which can be paused and played. The user is also able to change the playlist and adjust the volume. For the *Social panel*, the user may scroll through the notifications and delete those that he already checked. Interaction was designed so that any action can be carried out through all three means (visual, audio, gesture).

3.3.1. Visual Interaction

The visual interface provides visual information to the user and the touchscreen can let the user input information and perform actions on the active elements. On the *Drive panel*, the user may scroll the timeline for planning the trip. On the *Wellness panel* he may scroll through the playlists, change the volume by a vertical scroll and stop and play a song by a simple tap. On the *Social panel* he may scroll through the notifications, select one, and drag it for deletion.

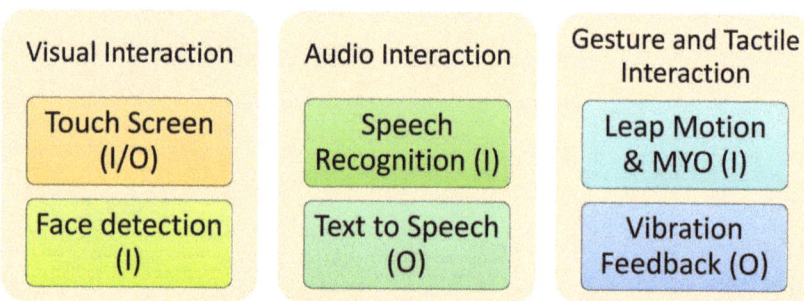

Figure 7. Multimodal natural interaction of the Smart Driving Interface (I is for input; O is for output).

The face detection for emotion, tiredness, and gaze direction are forms of passive interaction from the user's side. It means these are used for monitoring the user's state and enabling proactive action. We have experimented with smile as an indicator of approval for a suggestion made by the system, while a neutral face meant rejection of the proposal. An example is, when the user has a nervous demeanor or the pulse is high, the system will recommend calming music, to which he can agree by smiling.

3.3.2. Audio Interaction

The system receives auditory information through the Voice Recognition service of the Smart Watch. Voice commands need to follow the pattern specified by a grammar described with the help of the CMUSphinx Open Source Speech Recognition API (https://cmusphinx.github.io/doc/python/, accessed on 1 July 2021) [125,126]. There is a start phrase, a command, an object of interest, and a finish phrase. An example would be <please> <play> <the music> <thank you>. We should note that this command is available even if the *Wellness panel*, where the music is handled, is not currently the one selected, unlike with gesture interaction, where gestures affect only the current selected panel.

The text-to-speech component uses a voice synthesizer to read alerts produced by the system. We have experimented with sound and voice alerts and silent vibration feedback alerts. When an audio alert is being transmitted, the music is paused momentarily, and automatically played once the alert has finished reading.

3.3.3. Gesture Interaction

The Leap Motion device (https://developer.leapmotion.com/, accessed on 1 July 2021) can interpret midair gestures performed by the user by using an active IR scanner and algorithms to process the 3D data obtained (different shades of illumination indicate depth). It can discriminate between swipe left and right gestures, circular left and right gestures, a tap gesture, and a key press gesture, the last two having also a location attribute. It also knows the hand orientation, magnitude of openness (open palm vs. closed fist), and finger positions, and new gestures may be described for added complexity [127–129].

We experimented with different approaches to handling actions with the help of gestures and implemented a similar sequence of gestures that would resemble a grammar. For a starting gesture we chose an open hand, performed for at least 1 s. Then the user may swipe to reach the desired screen. Once the screen is selected using a tap, the user may perform a circular motion to scroll through the timeline, playlists, or notifications. Another tap will bring the timeline to the present moment for the *Driving panel*. The same action will enable deleting a notification on the *Social panel*, which is performed by a swipe. A tap on the *Wellness panel* will enable pausing the music by closing the hand and playing it by opening the hand. The volume will also be active for change by a circular motion to the left or right.

Vibrational feedback is performed by the Smart Watch on the user's hand and helps him notice that the Leap Motion device is waiting for a sequence of input gestures. It also serves as a confirmation of voice or gesture command approval by the system. We have also implemented vibration as a form of silent alert instead of pausing the music and reading the alert using the text-to-speech component.

We have also experimented with natural interaction through gestures using the MYO (https://developerblog.myo.com/, accessed on 1 July 2021) device, which reads muscular contractions through a Electromyography EMG interface and infers gestures. It knows the configuration of the hand because gestures such as a stop sign or a gun shooting gesture involve different muscular subgroups. It also knows the dynamics of the gestures using its inertial measurement unit composed of a three-axis gyroscope, accelerometer and magnetometer.

3.4. Autonomy and Proactive Interaction

By default, in the absence of interaction from the user's side, the system will constantly switch between the three panels every 20 s. If there is an event occurring in one of the three areas, driving, wellness, or social, then the respective panel will be displayed, along with a short message or an alert about the event.

The *Driving panel* events are related to newly detected hazards or approaching already detected hazards. The *Wellness panel* events are related to the detected tiredness or stress. The *Social panel* events are related to new incoming notifications or notifications that were postponed are now being shown.

3.4.1. Active Monitoring

An intelligent system built under the ubiquitous computing paradigm is reactive to the active form of interaction from the user side, as well as proactive, by monitoring the user, and acting in consequence. The user is thus actively monitored in term of pulse, emotion, alertness, and gaze direction. The car and the environment are also monitored in terms of speed, and natural conditions, respectively.

We selected and categorized eight indicators of the driving task quality, graphically represented in Figure 8. They are grouped into external overload (from monitoring the environment) and internal stress (from monitoring the user) factors, and stimulation that heightens or lowers the state of alertness. Using these eight parameters we infer if the user is either deprived or overloaded by stimuli, and if his internal state (caused by the external factors or by other internal or external problems) is optimal or there is a risk of tiredness or stress respectively.

3.4.2. Proactive Action

Safe driving depends on the alertness of the driver. We wish to keep an optimal state of alertness for the driver by balancing the external stimuli that we can control, such as the notifications from the *Social panel* and the music on the *Wellness panel*.

If there is a risk of the driver to feel asleep, the system will alert him on the *Wellness panel* with a blue heart or blue emoji and suggest an energetic playlist with a fast tempo. If there is a risk that the user is overloaded by stimuli that impair his focus on the driving task, the incoming notification on the *Social panel* will be momentarily queued and the system will alert him on the *Wellness panel* with a red heart or red emoji, suggesting a calm playlist.

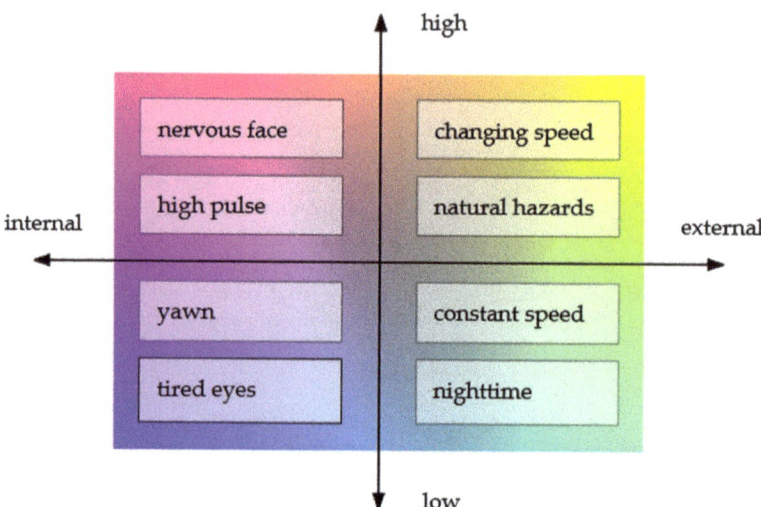

Figure 8. Factors affecting the quality and safety of driving.

4. Heuristic Evaluation of the IVIS

Various components of the proposed in-vehicle system have been subjected to usability studies, with encouraging results obtained from the users that participated in the experimental evaluation and answered the subsequent usability questionnaires. The now complete system incorporates the previously developed components [125,126,130,131], but employs several important extensions, such as predicting the weather and lighting conditions along the road and analyzing the face of the driver in terms of emotion, blinking and yawning.

In previous work [130,131] we described a component of the current system that assists the driver by automatically filtering, in an adaptive manner, messages and notifications from third party applications, that may distract the driver's attention while driving. We performed a usability study involving a sample of 75 users, selected from a pool of university personnel and students. The subsystem's level of usability and acceptability was appreciated by the users that expressed a high level of experience with mobile apps (Figure 9, details in [131]).

A thorough study regarding the adaptive positioning of the information displayed by the in-vehicle system onto the windshield, based on the automatic detection of the drivers' head orientation was previously performed in [126]. With the purpose of providing the driver with the information of interest, while focusing on the road at all times, we performed a usability testing experiment of the proposed visualization system, with a sample of 25 people, aged between 19 and 34 years, with almost 68% males. We found that users preferred a 20/25 cm sized display, positioned in the central area horizontally, combined with the middle/low area of the windshield vertically, and a black and white projector. Details on the entire usability study are presented in our work [125].

While user testing of components employed a scenario-based approach, the heuristic evaluation provides a distinct way of evaluation, which is complementary to user testing and focuses on identifying problems of the design, with respect to the formal guidelines [132]. At the core of a heuristic analysis lies the expertise of a human engineer [133]. In the current development phase, the in-vehicle system was subjected to heuristic evaluation holistically, in the sense that it was evaluated with respect to a set of usability heuristics by a set of three evaluators. They were introduced to the system and a typical usage scenario, by presenting them various steps that can be undertaken when using the system in order to perform some in-vehicle tasks, such as planning the departure time of a trip having knowledge of the future local weather and illumination conditions and possible hazards, getting a musical track recommendation based on their emotional/alertness status and the driving conditions, and checking their phone notifications that are updated only when it is considered safe to do so.

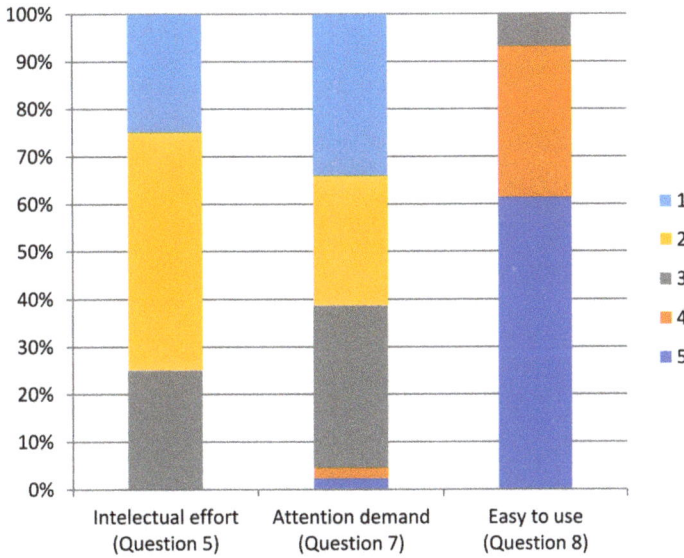

Figure 9. Results on the evaluation of the satisfaction degree of the users regarding the adaptive notification filtering component of the system [131] (percentage of respondents selecting an answer on the Likert scale, 1 for total disagreement, and 5 for total agreement with the questionnaire item; Q5. Interaction with the application requires a high intellectual effort, Q7. The information displayed was difficult to read, Q8. I think that the display responded quite quickly to the information updates).

The ergonomic criteria that were taken into consideration for evaluation are presented in Table 1, along with the performance of the system. We have compared our system to the guidelines about HMI design for automated driving systems, specified in [132]. We have selected 15 out of 20 guidelines that fit our type of system which does not handle automated driving. We described in Table 1 how our system complies or does not comply to these rules. These guidelines are following the ISO 9241 (Ergonomics of human-system interaction—Part 11: Usability: Definitions and concepts. https://www.iso.org/obp/ui/#iso:std:iso:9241:-11:ed-2:v1:en, accessed on 1 July 2021) standard which specifies that the usability measures "the extent to which a product can be used by specified users to achieve specific goals with effectiveness (can you achieve the goal?), efficiency (how many resources do you consume to achieve the goal?) and satisfaction (that is the quality of your experience when using the system to achieve your goals?) in a specified context of use".

Table 1. System evaluation for compliance with HMI guidelines for automated vehicles, as proposed in [132].

Human Machine Interface Evaluation Guidelines [132]	Compliance	Commentary
(2) "The system mode should be displayed continuously"	The system displays the inferred state of the driver, instead of showing the state of the system. If notifications are postponed, a pause sign is shown.	In case the system changes state, the interface just switches to either the driving or *Wellness panel*, where the alert is coming from.

Table 1. *Cont.*

Human Machine Interface Evaluation Guidelines [132]	Compliance	Commentary
(3) "System state changes should be effectively communicated"	Connection state is always on display. Tiredness alerts are actively communicated. Car and road alerts are also actively communicated.	Stress alerts are silent. Most of the changes take place in silence in order not to distract the driver.
(5) "HMI elements should be grouped together according to their function to support the perception of mode indicators"	The system groups information into three groups: — driving task-related, — wellness-related, — social-related	-
(7) "The visual interface should have a sufficient contrast in luminance and/or colour between foreground and background"	We chose white and full saturated colors with a 50% lightness, on a dark background	-
(8) "Texts (e.g., font types and size of characters) and symbols should be easily readable from the permitted seating position"	-	The text of the notifications on the *Social panel* is not very short, nor is the font large enough. The system displays notifications when it is considered relatively safe for the driver to check them.
(9) "Commonly accepted or standardized symbols should be used to communicate the automation mode. Use of non-standard symbols should be supplemented by additional text explanations or vocal phrase/s"	We chose colored icons to express the state of the driver, the car, and the road. With respect to the state of the driver *red* means stress, while *blue* means tiredness. With respect to the information gathered from the car by means of the OBDII interface, we communicate warnings. The weather is shown using conventional icons, while for illumination we use color codes: blue shades for the sky and orange for sunglare hazards.	-
(10) "The semantic of a message should be in accordance with its urgency"	Semantics is communicated through text and color.	-
(12) "Text messages should be as short as possible"	-	The text body of smart phone notifications is fully displayed.
(13) Not more than five colours should be consistently used to code system states (excluding white and black)	Connection state of the system from the server and other components is coded with *green* for connected and *red* for disconnected.	We chose to express states for the pulse, tiredness/stress, music through the spectrum of colors between *red* and *blue* included.
(14) "The colours used to communicate system states should be in accordance with common conventions and stereotypes"	*Red* is for stress, high pulse, and also for energetic music, hazards. *Green* for optimal physiological state. *Blue* is for tiredness, low pulse, calming music.	We do use all colors between *red* and *blue* because the spectrum is mapped to each possible value/meaning. Pulse according to continuous intervals, state of alertness fades to green after a red or blue alert.
(16) "Auditory output should raise the attention of the driver without startling her/him or causing pain"	We consider as high priority the pleasantness of multimodal outputs of the system.	-

Table 1. Cont.

Human Machine Interface Evaluation Guidelines [132]	Compliance	Commentary
(17) "Auditory and vibrotactile output should be adapted to the urgency of the message"	The system only raises alerts for vehicle-related problems, environmental hazards, detected driver tiredness and low environmental stimulation that may lead to sleep. These alerts are actively communicated.	Stress alerts would also be high on the priority list, but we chose not to distract the driver even more than he already is. We just show the *Wellness panel* with a red facial icon, and silently recommend calming music. He may smile for agreement.
(18) "High-priority messages should be multimodal"	The system uses both audio output and vibration output.	-
(19) "Warning messages should orient the user towards the source of danger"	Every time an alert is shown, the system switches to the specific information panel.	-

The current heuristic testing of the in-vehicle system is not supposed to replace usability testing with heterogeneous participant samples, but merely to provide a basis for further thorough evaluation of the in-vehicle system. By means of the current evaluation of the system, we are able to devise further plans for empirical testing, in order to identify issues that may have been overlooked at this point. A critical evaluation of the system with respect to strengths, weaknesses, opportunities and threats is detailed in Table 2.

Table 2. SWOT analysis of the proposed in-vehicle system.

Strengths	Weaknesses
• The system integrates in an innovative manner data from smart sensors, processed information, capabilities of natural interaction, autonomous control. • We promote driving safety through complementary means to the car's automation systems that are useful in any type of car. • Our interface does respect some important design guidelines for in-vehicle HMI. • The costs of the prototype are small, considering that we only used devices that are affordable, and the software relies only on freely available, open-source libraries, apart from the original software especially written for the system.	• The system relies heavily on internet service, while for the location it relies on mobile services • The system described in the paper, while fully functional, is a prototype. It is not yet integrated with other expected components, such as crowd sourced navigation services or landmark search (as compared to Google Maps) • The system uses open-source software libraries that were not specially designed for vehicles, which were adapted to suit our current needs.
Opportunities	Threats
• Most of the devices employed by the proposed (except the OBDII interface) can easily be installed in any car. • The system offers diverse functionalities at a low cost, compared to those available in high end car models. • The system is most probably a concept totally new and interesting for most people, especially in today's digital society.	• On the part of the research team: the team's resources were limited by people number, time, funds. • On the part of the intended users of the system: some drivers are not technology enthusiasts, or they may have privacy issues or dislike being monitored altogether. • On the part of the vehicles involved: other cars may be able to connect to services by peer-to-peer networks. Security issues should be seriously taken into account, when developing the system for the public.

5. Conclusions

We tackled the problem of designing a useful, safe, supportive, and comfortable environment for the driver and passengers in the "digital car-sphere", composed of the vehicle, various sensors, personal mobile and/or wearable devices. In this respect, we proposed a system that departs from the conventional in-vehicle interface, in the sense that the entire in-vehicle digital space is tackled as an interface for the intelligent digital environment. The system relies on various onboard or outboard sensors (speed, pulse, CMOS etc.), devices (smart phone, smart watch, tablet etc.), and systems (laptop with LeapMotion, MYO etc.), which gather, integrate, and process data, providing in return relevant information and proactive interaction with the driver, passengers or other systems.

Intending to identify the challenging issues faced by the driver, we aimed to properly define the problem of integrating smart driving interface components which process different categories of information, relying on Design Thinking [72] principles. Consequently, we proposed suitable solutions by focusing on designing an in-vehicle GUI that has to be useful, effective and which bring enjoyable user experience. Moreover, as only the relevant information is presented to the driver at a time we argue that this will have a positive effect on the safety of driving. From an implementation point-of-view we relied heavily on open source or free libraries/APIs/SDKs such as MapBox API, SunCalc library, OpenWeather API, YouTube Player API, Beyond Reality Face SDK, FaceApi, CMUSphinx Open-Source Speech Recognition, LeapMotion SDK, etc.

A typical use case scenario for the system develops as the driver engages the in-vehicle activities, while the intelligent in-vehicle system tracks its actions, supporting adaptation to the ongoing tasks and enhancing the driver experience. The system setup considers four main components: data, information, interaction, and autonomy. The data consists of raw values collected together from the entire driving environment (in- or out-vehicle), which are further classified into several categories, namely Driving, Wellness, Social. The multi-modal interface of the system incorporates different types of interactions tailored to the user's perceptual senses and to the driving context. The Intelligent digital environment provides continuous feedback as a consequence of driver/passenger actions and other in- or -out vehicle information inferred from the collected data.

Previous efforts in the development of the system were directed towards incorporating audio and gestural interaction within an in-vehicle system meant to foster driver's attentiveness, coupled with investigating the use of augmented reality in order to enhance the data visualization inside a smart vehicle [125,126,130,131]. The usability of the referred components was assessed by user-based studies, in a controlled laboratory environment meant to simulate the in-vehicle space. The system proposed in this paper incorporates and extends those components, offering a functional complex in-vehicle system, which uses multi-modal interaction that assists the user in the driving task. The driver, who is at the center of the system, is interacting with the system by means of various modalities. Some interactions are voluntary, such as touch input, voice, and gesture commands. On the other hand, the interaction may also be involuntary, in the form of the system reading the driver's facial expressions, hand gestures, or inferring the emotional state (for example inferring stress from pulse data, and tiredness from yawning and abnormal blinking).

The system was evaluated according to guidelines especially defined for the evaluation of human-machine interfaces in the context of automated vehicles. A SWOT analysis is also presented, revealing strengths and weaknesses of the current in-vehicle system proposal. Thorough usability testing will be employed as future work, as soon as pandemic related conditions will allow for selecting a relevant sample of users to test the system in the laboratory setting. Further development of the system will be targeted to secure the system against unwanted access by third parties.

Author Contributions: Conceptualization, E.B. (Emanuela Bran), E.B. (Elena Bautu), C.M.P., D.F.S. and D.M.P.; methodology, E.B. (Emanuela Bran), E.B. (Elena Bautu) and C.M.P.; software, E.B. (Emanuela Bran), E.B. (Elena Bautu), C.M.P., D.F.S. and D.M.P.; validation, E.B. (Emanuela Bran), and D.M.P.; formal analysis, E.B. (Emanuela Bran), C.M.P. and E.B. (Elena Bautu); investigation, E.B. (Emanuela Bran) and E.B. (Elena Bautu); resources, D.M.P.; data curation, E.B. (Emanuela Bran); writing—original draft preparation, E.B. (Emanuela Bran), E.B. (Elena Bautu), C.M.P., D.F.S. and D.M.P.; writing—review and editing, E.B. (Emanuela Bran) and E.B. (Elena Bautu); visualization, E.B. (Emanuela Bran); supervision, D.M.P.; project administration, D.M.P.; funding acquisition, D.M.P. All authors have read and agreed to the published version of the manuscript.

Funding: This research was funded by the Romanian Ministry of Research and Innovation, CCCDI-UEFISCDI, project number PN-III-P1-1.2-PCCDI-2017-0917, contract no. 21PCCDI/2018, within PNCDI III, project P2, "Efficient communications based on smart devices for in-car augmented reality interactive applications".

Acknowledgments: We thank the anonymous reviewers for their comments and insights that significantly improved our paper.

Conflicts of Interest: The authors declare no conflict of interest.

References

1. Arena, F.; Pau, G.; Severino, A. An overview on the current status and future perspectives of smart cars. *Infrastructures* **2020**, *5*, 53. [CrossRef]
2. Trubia, S.; Severino, A.; Curto, S.; Arena, F.; Pau, G. Smart Roads: An Overview of What Future Mobility Will Look Like. *Infrastructures* **2020**, *5*, 107. [CrossRef]
3. Seiberth, G.; Gründinger, W. Data-driven business models in connected cars, mobility services & beyond. *BVDW Res.* **2018**, *1*, 18.
4. Rijcken, C. Rainforests of wearables and insideables. In *Pharmaceutical Care in Digital Revolution*; Academic Press: Cambridge, MA, USA, 2019; pp. 107–117.
5. Gheran, B.F.; Vatavu, R.D. From controls on the steering wheel to controls on the finger: Using smart rings for in-vehicle interactions. In *Companion Publication of the 2020 ACM Designing Interactive Systems Conference*; Massachusetts Institute of Technology: Cambridge, MA, USA, 2020; pp. 299–304.
6. Pelliccione, P.; Knauss, E.; Ågren, S.M.; Heldal, R.; Bergenhem, C.; Vinel, A.; Brunnegård, O. Beyond connected cars: A systems of systems perspective. *Sci. Comput. Program.* **2020**, *191*, 102414. [CrossRef]
7. Telang, S.; Chel, A.; Nemade, A.; Kaushik, G. Intelligent Transport System for a Smart City. In *Security and Privacy Applications for Smart City Development*; Springer: Cham, 2021; pp. 171–187.
8. Eiza, M.H.; Cao, Y.; Xu, L. *Toward Sustainable and Economic Smart Mobility: Shaping the Future of Smart Cities*; WSPC: Casper, WY, USA, 2020.
9. Ahram, T.; Karwowski, W.; Vergnano, A.; Leali, F.; Taiar, R. Intelligent Human Systems Integration. In Proceedings of the 3rd International Conference on Intelligent Human Systems Integration (IHSI 2020): Integrating People and Intelligent Systems, Modena, Italy, 19–21 February 2020; Springer Nature: Cham, Switzerland, 2020.
10. Nischak, F.; Hanelt, A. Ecosystem Change in the Era of Digital Innovation–A Longitudinal Analysis and Visualization of the Automotive Ecosystem. In Proceedings of the ICIS 2019 Proceedings, Munich, Germany, 15–18 December 2019; ISBN 978-0-9966831-9-7.
11. Uddin, H.; Gibson, M.; Safdar, G.A.; Kalsoom, T.; Ramzan, N.; Ur-Rehman, M.; Imran, M.A. IoT for 5G/B5G applications in smart homes, smart cities, wearables and connected cars. In Proceedings of the 2019 IEEE 24th International Workshop on Computer Aided Modeling and Design of Communication Links and Networks (CAMAD), Limassol, Cyprus, 23 June 2019; pp. 1–5.
12. Kuoch, S.K.; Nowakowski, C.; Hottelart, K.; Reilhac, P.; Escrieut, P. Designing an Intuitive Driving Experience in a Digital World. *Automot. Eng.* **2018**. preprint. [CrossRef]
13. Budaker, B.; Geiger, M.; Fernandes, K. Development of smart interior systems for connected cars. In *Internationales Stuttgarter Symposium*; Springer Vieweg: Wiesbaden, Germany, 2018; pp. 1265–1276.
14. Perelló, J.R.; García, A. A case study of cooperative design on integrated smart-car systems: Assessing drivers' experience. In *International Conference on Cooperative Design, Visualization and Engineering*; Springer: Cham, Switzerland, 2017; pp. 202–206.
15. Broström, R.; Engström, J.; Agnvall, A.; Markkula, G. Towards the next generation intelligent driver information system (IDIS): The Volvo car interaction manager concept. In Proceedings of the 2006 ITS World Congress, London, UK, 8–12 October 2006; Volume 32.
16. Han, J.; Kim, H.; Heo, S.; Lee, N.; Kang, D.; Oh, B.; Kim, K.; Yoon, W.; Byun, J.; Kim, D. GS1 Connected Car: An Integrated Vehicle Information Platform and Its Ecosystem for Connected Car Services based on GS1 Standards. In Proceedings of the 2018 IEEE Intelligent Vehicles Symposium (IV), Changshu, China, 26–30 June 2018; pp. 367–374.
17. Lipson, H.; Kurman, M. *Driverless: Intelligent Cars and the Road ahead*; Mit Press: Cambridge, MA, USA, 2016.

18. Großwindhager, B.; Rupp, A.; Tappler, M.; Tranninger, M.; Weiser, S.; Aichernig, B.K.; Boano, C.A.; Horn, M.; Kubin, G.; Mangard, S.; et al. Dependable internet of things for networked cars. *Int. J. Comput.* **2017**, *16*, 226–237. [CrossRef]
19. Vörös, F.; Tompos, Z.; Kovács, B. Examination of car navigation systems and UX designs–suggestion for a new interface. *Proc. Int. Cartogr. Assoc* **2019**, *2*, 139. [CrossRef]
20. Kazmi, S.A.; Dang, T.N.; Yaqoob, I.; Ndikumana, A.; Ahmed, E.; Hussain, R.; Hong, C.S. Infotainment enabled smart cars: A joint communication, caching, and computation approach. *IEEE Trans. Veh. Technol.* **2019**, *68*, 8408–8420. [CrossRef]
21. Pandit, S.; Fitzek, F.H.; Redana, S. Demonstration of 5G connected cars. In Proceedings of the 2017 14th IEEE Annual Consumer Communications & Networking Conference (CCNC), Las Vegas, NV, USA, 8–11 January 2017; pp. 605–606.
22. Giust, F.; Sciancalepore, V.; Sabella, D.; Filippou, M.C.; Mangiante, S.; Featherstone, W.; Munaretto, D. Multi-access edge computing: The driver behind the wheel of 5G-connected cars. *IEEE Commun. Stand. Mag.* **2018**, *2*, 66–73. [CrossRef]
23. Uhlir, D.; Sedlacek, P.; Hosek, J. Practial overview of commercial connected cars systems in Europe. In Proceedings of the 2017 9th International Congress on Ultra Modern Telecommunications and Control Systems and Workshops (ICUMT), Munich, Germany, 6–8 November 2017; pp. 436–444.
24. Marosi, A.C.; Lovas, R.; Kisari, Á.; Simonyi, E. January. A novel IoT platform for the era of connected cars. In Proceedings of the 2018 IEEE International Conference on Future IoT Technologies (Future IoT), Eger, Hungary, 18–19 January 2018; pp. 1–11.
25. Jiang, T.; Fang, H.; Wang, H. Blockchain-Based Internet of Vehicles: Distributed Network Architecture and Performance Analysis. *IEEE Internet Things J.* **2018**, *6*, 4640–4649. [CrossRef]
26. Xu, W.; Zhou, H.; Cheng, N.; Lyu, F.; Shi, W.; Chen, J.; Shen, X. Internet of vehicles in big data era. *IEEE J. Autom. Sin.* **2017**, *5*, 19–35. [CrossRef]
27. Zhdanenko, O.; Liu, J.; Torre, R.; Mudriievskiy, S.; Salah, H.; Nguyen, G.T.; Fitzek, H.F. Demonstration of mobile edge cloud for 5g connected cars. In Proceedings of the 2019 16th IEEE Annual Consumer Communications & Networking Con-ference (CCNC), Las Vegas, NV, USA, 11–14 January 2019; pp. 1–2.
28. Ma, Z.; Zhang, J.; Guo, Y.; Liu, Y.; Liu, X.; He, W. An Efficient Decentralized Key Management Mechanism for VANET With Blockchain. *IEEE Trans. Veh. Technol.* **2020**, *69*, 5836–5849. [CrossRef]
29. Ayaz, F.; Sheng, Z.; Tian, D.; Leung, V.C. Blockchain-enabled security and privacy for Internet-of-Vehicles. In *Internet of Vehicles and its Applications in Autonomous Driving*; Springer: Cham, Switzerland, 2021; pp. 123–148.
30. Hernandez-Oregon, G.; Rivero-Angeles, M.E.; Chimal-Eguía, J.C.; Campos-Fentanes, A.; Jimenez-Gallardo, J.G.; Es-tevez-Alva, U.O.; Juarez-Gonzalez, O.; Rosas-Calderon, P.O.; Sandoval-Reyes, S.; Menchaca-Mendez, R. Performance analysis of V2V and V2I LiFi communication systems in traffic lights. *Wirel. Commun. Mobile Comput.* **2019**, *2019*, 4279683. [CrossRef]
31. Spahiu, C.S.; Stanescu, L.; Brezovan, M.; Petcusin, F. LiFi Technology Feasibility Study for Car-2-Car Communication. In Proceedings of the 21th International Carpathian Control Conference (ICCC), High Tatras, Slovakia, 27–29 October 2020; pp. 1–5.
32. Burkacky, O.; Deichmann, J.; Doll, G.; Knochenhauer, C. *Rethinking Car Software and Electronics Architecture*; McKinsey & Co.: New York, NY, USA, 2019.
33. Yang, G.; Ahmed, M.; Gaweesh, S.; Adomah, E. Connected vehicle real-time traveler information messages for freeway speed harmonization under adverse weather conditions: Trajectory level analysis using driving simulator. *Accid. Anal. Prev.* **2020**, *146*, 105707. [CrossRef]
34. Jagielski, M.; Jones, N.; Lin, C.W.; Nita-Rotaru, C.; Shiraishi, S. Threat detection for collaborative adaptive cruise control in connected cars. In Proceedings of the 11th ACM Conference on Security & Privacy in Wireless and Mobile Networks, Stockholm, Sweden, 18–20 June 2018; pp. 184–189.
35. Lee, E.-K.; Gerla, M.; Pau, G.; Lee, U.; Lim, J.-H. Internet of Vehicles: From intelligent grid to autonomous cars and vehicular fogs. *Int. J. Distrib. Sens. Netw.* **2016**, *12*, 1550147716665500. [CrossRef]
36. Gerla, M.; Lee, E.-K.; Pau, G.; Lee, U. Internet of vehicles: From intelligent grid to autonomous cars and vehicular clouds. In *IEEE World Forum on Internet of Things (WF-IoT)*; IEEE: Piscataway, NJ, USA, 2014; pp. 241–246. [CrossRef]
37. Bosler, M.; Jud, C.; Herzwurm, G. Platforms and Ecosystems for Connected Car Services. In Proceedings of the 9th International Workshop Software Ecosystem IWSECO, Espoo, Finland, 29 November 2017; pp. 16–27.
38. Zhou, H.; Xu, W.; Chen, J.; Wang, W. Evolutionary V2X Technologies toward the Internet of Vehicles: Challenges and Opportunities. *Proc. IEEE* **2020**, *108*, 308–323. [CrossRef]
39. Mirnig, N.; Perterer, N.; Stollnberger, G.; Tscheligi, M. Three strategies for autonomous car-to-pedestrian communication: A survival guide. In Proceedings of the Companion of the 2017 ACM/IEEE International Conference on Human-Robot Interaction, Vienna, Austria, 6–9 March 2017; pp. 209–210.
40. Bai, H.; Shen, J.; Wei, L.; Feng, Z. Accelerated Lane-Changing Trajectory Planning of Automated Vehicles with Vehicle-to-Vehicle Collaboration. *J. Adv. Transp.* **2017**, *2017*, 8132769. [CrossRef]
41. Hock, P.; Benedikter, S.; Gugenheimer, J.; Rukzio, E. Carvr: Enabling in-car virtual reality entertainment. In Proceedings of the 2017 CHI Conference on Human Factors in Computing Systems, Denver, CO, USA, 6–11 May 2017; pp. 4034–4044.
42. Malinverno, M.; Mangues-Bafalluy, J.; Casetti, C.E.; Chiasserini, C.F.; Requena-Esteso, M.; Baranda, J. An Edge-Based Framework for Enhanced Road Safety of Connected Cars. *IEEE Access* **2020**, *8*, 58018–58031. [CrossRef]
43. Bierzynski, K.; Escobar, A.; Eberl, M. Cloud, fog and edge: Cooperation for the future? In Proceedings of the 2017 Second International Conference on Fog and Mobile Edge Computing (FMEC), Valencia, Spain, 8–11 May 2017; pp. 62–67.

44. Ghosh, S.; Mukherjee, A.; Ghosh, S.K.; Buyya, R. Mobi-iost: Mobility-aware cloud-fog-edge-iot collaborative framework for time-critical applications. *IEEE Trans. Netw. Sci. Eng.* **2019**, *7*, 2271–2285. [CrossRef]
45. Vallati, M. Centralised Versus Decentralised Traffic Optimisation of Urban Road Networks: A Simulation Study. In Proceedings of the 2020 IEEE 5th International Conference on Intelligent Transportation Engineering (ICITE), Beijing, China, 11–13 September 2020; pp. 319–325. [CrossRef]
46. Duan, L.; Wei, Y.; Zhang, J.; Xia, Y. Centralized and decentralized autonomous dispatching strategy for dynamic autonomous taxi operation in hybrid request mode. *Transp. Res. Part C Emerg. Technol.* **2020**, *111*, 397–420. [CrossRef]
47. Olaverri-Monreal, C.; Lehsing, C.; Trubswetter, N.; Schepp, C.A.; Bengler, K. In-vehicle displays: Driving information prioritization and visualization. In Proceedings of the 2013 IEEE Intelligent Vehicles Symposium (IV), Gold Coast, Australia, 23–26 June 2013; pp. 660–665. [CrossRef]
48. Siems-Anderson, A.R.; Walker, C.L.; Wiener, G.; Mahoney, W.P., III; Haupt, S.E. An adaptive big data weather system for surface transportation. *Transp. Res. Interdiscip. Perspect.* **2019**, *3*, 100071. [CrossRef]
49. Kamoun, F.; Chaabani, H.; Outay, F.; Yasar, A.-U. A Survey of Approaches for Estimating Meteorological Visibility Distance under Foggy Weather Conditions. *IGI Glob.* **2020**, 65–92. [CrossRef]
50. Wang, K.; Zhang, W.; Feng, Z.; Yu, H.; Wang, C. Reasonable driving speed limits based on recognition time in a dynamic low-visibility environment related to fog—A driving simulator study. *Accid. Anal. Prev.* **2021**, *154*, 106060. [CrossRef]
51. Hold-Geoffroy, Y.; Sunkavalli, K.; Hadap, S.; Gambaretto, E.; Lalonde, J.F. Deep outdoor illumination estimation. In Proceedings of the IEEE Conference on Computer Vision and Pattern Recognition, Honolulu, HI, USA, 21–26 July 2017; pp. 7312–7321.
52. Wood, J.M. Nighttime driving: Visual, lighting and visibility challenges. *Ophthalmic Physiol. Opt.* **2020**, *40*, 187–201. [CrossRef]
53. Pegin, P.; Sitnichuk, E. The Effect of Sun Glare: Concept, Characteristics, Classification. *Transp. Res. Procedia* **2017**, *20*, 474–479. [CrossRef]
54. Li, X.; Cai, B.Y.; Qiu, W.; Zhao, J.; Ratti, C. A novel method for predicting and mapping the occurrence of sun glare using Google Street View. *Transp. Res. Part C Emerg. Technol.* **2019**, *106*, 132–144. [CrossRef]
55. Glaser, S.; Mammar, S.; Dakhlallah, D. Lateral wind force and torque estimation for a driving assistance. *IFAC Proc. Vol.* **2008**, *41*, 5688–5693. [CrossRef]
56. Li, Y.; Xing, L.; Wang, W.; Wang, H.; Dong, C.; Liu, S. Evaluating impacts of different longitudinal driver as-sistance systems on reducing multi-vehicle rear-end crashes during small-scale inclement weather. *Accid. Anal. Prev.* **2017**, *107*, 63–76. [CrossRef]
57. Darwish, T.S.; Bakar, K.A. Fog based intelligent transportation big data analytics in the internet of vehicles environment: Motivations, architecture, challenges, and critical issues. *IEEE Access* **2018**, *6*, 15679–15701. [CrossRef]
58. Hirz, M.; Walzel, B. Sensor and object recognition technologies for self-driving cars. *Comput. Des. Appl.* **2018**, *15*, 501–508. [CrossRef]
59. Krumm, J. *Ubiquitous Computing Fundamentals*; CRC Press: Boca Raton, FL, USA, 2018.
60. Brush, A.B. Ubiquitous Computing Field Studies. In *Ubiquitous Computing Fundamentals*; Chapman and Hall/CRC: London, UK, 2018; pp. 175–216.
61. Saganowski, S.; Kazienko, P.; Dzieżyc, M.; Jakimów, P.; Komoszyńska, J.; Michalska, W.; Dutkowiak, A.; Polak, A.; Dziadek, A.; Ujma, M. Consumer wearables and affective computing for wellbeing support. *arXiv* **2020**, preprint. arXiv:2005.00093.
62. El-Gayar, O.F.; Ambati, L.S.; Nawar, N. Wearables, artificial intelligence, and the future of healthcare. In *AI and Big Data's Potential for Disruptive Innovation*; IGI Global: Hershey, PA, USA, 2020; pp. 104–129.
63. Hicks, J.L.; Althoff, T.; Sosic, R.; Kuhar, P.; Bostjancic, B.; King, A.C.; Leskovec, J.; Delp, S.L. Best practices for analyzing large-scale health data from wearables and smartphone apps. *NPJ Digit. Med.* **2019**, *2*, 45. [CrossRef] [PubMed]
64. Lou, M.; Abdalla, I.; Zhu, M.; Wei, X.; Yu, J.; Li, Z.; Ding, B. Highly Wearable, Breathable, and Washable Sensing Textile for Human Motion and Pulse Monitoring. *ACS Appl. Mater. Interfaces* **2020**, *12*, 19965–19973. [CrossRef] [PubMed]
65. Liang, R.H.; Yang, S.Y.; Chen, B.Y. Indexmo: Exploring finger-worn RFID motion tracking for activity recognition on tagged objects. In Proceedings of the 23rd International Symposium on Wearable Computers, London, UK, 19–13 September; pp. 129–134.
66. Bandodkar, A.J.; Jia, W.; Yardımcı, C.; Wang, X.; Ramirez, J.; Wang, J. Tattoo-based noninvasive glucose mon-itoring: A proof-of-concept study. *Anal. Chem.* **2015**, *87*, 394–398. [CrossRef]
67. Andrew, T.L. The Future of Smart Textiles: User Interfaces and Health Monitors. *Matter* **2020**, *2*, 794–795. [CrossRef]
68. Kurasawa, S.; Ishizawa, H.; Fujimoto, K.; Chino, S.; Koyama, S. Development of Smart Textiles for Self-Monitoring Blood Glucose by Using Optical Fiber Sensor. *J. Fiber Sci. Technol.* **2020**, *76*, 104–112. [CrossRef]
69. Zhou, Z.; Padgett, S.; Cai, Z.; Conta, G.; Wu, Y.; He, Q.; Zhang, S.; Sun, C.; Liu, J.; Fan, E.; et al. Single-layered ultra-soft washable smart textiles for all-around ballistocardiograph, respiration, and posture monitoring during sleep. *Biosens. Bioelectron.* **2020**, *155*, 112064. [CrossRef]
70. Koyama, S.; Sakaguchi, A.; Ishizawa, H.; Yasue, K.; Oshiro, H.; Kimura, H. Vital Sign Measurement Using Covered FBG Sensor Embedded into Knitted Fabric for Smart Textile. *J. Fiber Sci. Technol.* **2017**, *73*, 300–308. [CrossRef]
71. Sinnapolu, G.; Alawneh, S. Integrating wearables with cloud-based communication for health monitoring and emergency assistance. *Internet Things* **2018**, *1–2*, 40–54. [CrossRef]
72. Betancourt Diaz, N.R. Wearables, Big Data and Design Thinking: Perspectives from the Wellbeing Industry. Available online: https://www.politesi.polimi.it/handle/10589/139431 (accessed on 1 July 2021).

73. Lin, F.-R.; Windasari, N.A. Continued use of wearables for wellbeing with a cultural probe. *Serv. Ind. J.* **2018**, *39*, 1140–1166. [CrossRef]
74. Persson, N.-K.; Martinez, J.G.; Zhong, Y.; Maziz, A.; Jager, E.W.H. Actuating Textiles: Next Generation of Smart Textiles. *Adv. Mater. Technol.* **2018**, *3*, 1700397. [CrossRef]
75. Kongahage, D.; Foroughi, J. Actuator Materials: Review on Recent Advances and Future Outlook for Smart Textiles. *Fibers* **2019**, *7*, 21. [CrossRef]
76. Rayes, A.; Salam, S. The things in iot: Sensors and actuators. In *Internet of Things from Hype to Reality*; Springer: Cham, Switzerland, 2017; pp. 57–77.
77. Kazeem, O.O.; Akintade, O.O.; Kehinde, L.O. Comparative study of communication interfaces for sensors and actuators in the cloud of internet of things. *Int. J. Internet Things* **2020**, *6*, 9–13.
78. Pawlowski, E.; Pawlowski, K.; Trzcielinska, J.; Trzcielinski, S. Designing and management of intelligent, autonomous environment (IAE): The research framework. In Proceedings of the International Conference on Human Systems Engineering and Design: Future Trends and Applications, Pula, Croatia, 22–24 September 2020; Springer: Cham, Switzerland, 2020; pp. 381–386.
79. Takayama, L. The motivations of ubiquitous computing: Revisiting the ideas behind and beyond the prototypes. *Pers. Ubiquitous Comput.* **2017**, *21*, 557–569. [CrossRef]
80. Ravenswaaij-Arts, C.M.; Kollee, L.A.; Hopman, J.C.; Stoelinga, G.B.; van Geijn, H.P. Heart rate variability. *Ann. Intern. Med.* **1993**, *118*, 436–447. [CrossRef]
81. Ranjan, Y.; Rashid, Z.; Stewart, C.; Conde, P.; Begale, M.; Verbeeck, D.; Boettcher, S.; Dobson, R.; Folarin, A.; Hyve, T.; et al. RADAR-Base: Open Source Mobile Health Platform for Collecting, Monitoring, and Analyzing Data Using Sensors, Wearables, and Mobile Devices. *JMIR mHealth uHealth* **2019**, *7*, e11734. [CrossRef]
82. Fakhrhosseini, S.M.; Jeon, M. How do angry drivers respond to emotional music? A comprehensive perspective on assessing emotion. *J. Multimodal User Interfaces* **2019**, *13*, 137–150. [CrossRef]
83. Abdi, L.; Ben Abdallah, F.; Meddeb, A. In-Vehicle Augmented Reality Traffic Information System: A New Type of Communication Between Driver and Vehicle. *Procedia Comput. Sci.* **2015**, *73*, 242–249. [CrossRef]
84. Abdi, L.; Meddeb, A. Driver information system: A combination of augmented reality, deep learning and vehicular Ad-hoc networks. *Multimed. Tools Appl.* **2017**, *77*, 14673–14703. [CrossRef]
85. Schipor, O.A.; Vatavu, R.D. Towards Interactions with Augmented Reality Systems in Hyper-Connected Cars. *EICS Workshops* **2019**, *2503*, 76–82.
86. Vögel, H.J.; Süß, C.; Hubregtsen, T.; André, E.; Schuller, B.; Härri, J.; Conradt, J.; Adi, A.; Zadorojniy, A.; Terken, J.; et al. Emotion-awareness for intelligent vehicle assistants: A research agenda. In Proceedings of the 2018 IEEE/ACM 1st International Workshop on Software Engineering for AI in Autonomous Systems (SEFAIAS), Gothenburg, Sweden, 28 May 2018; 2018; pp. 11–15.
87. Birek, L.; Grzywaczewski, A.; Iqbal, R.; Doctor, F.; Chang, V. A novel Big Data analytics and intelligent technique to predict driver's intent. *Comput. Ind.* **2018**, *99*, 226–240. [CrossRef]
88. Michalke, T.; Gepperth, A.; Schneider, M.; Fritsch, J.; Goerick, C. Towards a Human-like Vision System for Resource-Constrained Intelligent Cars. In Proceedings of the International Conference on Computer Vision Systems, Berlin, Germany, 11–14 March 2007. [CrossRef]
89. Davidsson, S.; Alm, H. Context adaptable driver information–Or, what do whom need and want when? *Appl. Ergon.* **2014**, *45*, 994–1002. [CrossRef]
90. McStay, A. *Emotional AI: The Rise of Empathic Media*; Sage: Thousand Oaks, CA, USA, 2018. [CrossRef]
91. McStay, A. Emotional AI, soft biometrics and the surveillance of emotional life: An unusual consensus on privacy. *Big Data Soc.* **2020**, *7*, 2053951720904386. [CrossRef]
92. Braun, M.; Schubert, J.; Pfleging, B.; Alt, F. Improving Driver Emotions with Affective Strategies. *Multimodal Technol. Interact.* **2019**, *3*, 21. [CrossRef]
93. Oehl, M.; Ihme, K.; Pape, A.-A.; Vukelić, M.; Braun, M. Affective Use Cases for Empathic Vehicles in Highly Automated Driving: Results of an Expert Workshop. In *International Conference on Human-Computer Interaction*; Springer: Cham, Switzerland, 2020; pp. 89–100. [CrossRef]
94. Frison, A.K.; Wintersberger, P.; Riener, A.; Schartmüller, C. Driving Hotzenplotz: A hybrid interface for vehicle control aiming to maximize pleasure in highway driving. In Proceedings of the 9th International Conference on Automotive User Interfaces and Interactive Vehicular Applications, Oldenburg, Germany, 24–27 September 2017; pp. 236–244.
95. Caon, M.; Demierre, M.; Abou Khaled, O.; Mugellini, E.; Delaigue, P. February. Enriching the user experience of a connected car with quantified self. In *International Conference on Intelligent Human Systems Integration*; Springer: Cham, Switzerland, 2020; pp. 66–72.
96. Giraldi, L. The Future of User Experience Design in the Interior of Autonomous Car Driven by AI. In *International Conference on Intelligent Human Systems Integration*; Springer: Cham, Switzerland, 2020; pp. 46–51.
97. Lindgren, T.; Fors, V.; Pink, S.; Bergquist, M.; Berg, M. On the way to anticipated car UX. In Proceedings of the 10th Nordic Conference on Human-Computer Interaction, Oslo, Norway, 29 September–3 October 2018; pp. 494–504.
98. Basu, C.; Yang, Q.; Hungerman, D.; Sinahal, M.; Draqan, A.D. Do you want your autonomous car to drive like you? In Proceedings of the 2017 12th ACM/IEEE International Conference on Human-Robot Interaction, Vienna, Austria, 6–9 March 2017; pp. 417–425.

99. Paredes, P.E.; Balters, S.; Qian, K.; Murnane, E.L.; Ordóñez, F.; Ju, W.; Landay, J.A. Driving with the fishes: Towards calming and mindful virtual reality experiences for the car. *Proc. ACM Interact. Mob. Wearable Ubiquitous Technol.* **2018**, *2*, 1–21. [CrossRef]
100. Silva, F.; Analide, C. Ubiquitous driving and community knowledge. *J. Ambient. Intell. Humaniz. Comput.* **2017**, *8*, 157–166. [CrossRef]
101. Oehl, M.; Ihme, K.; Bosch, E.; Pape, A.A.; Vukelić, M.; Braun, M. Emotions in the age of automated driving-developing use cases for empathic cars. In *Mensch und Computer 2019-Workshopband*; Gesellschaft für Informatik: Bonn, Germany, 2019.
102. Fonsalas, F. Holistic HMI Architecture for Adaptive and Predictive Car Interiors. In *Electronic Components and Systems for Automotive Applications*; Springer: Cham, Switzerland, 2019; pp. 217–227.
103. Neuhaus, R.; Laschke, M.; Theofanou-Fülbier, D.; Hassenzahl, M.; Sadeghian, S. Exploring the im-pact of transparency on the interaction with an in-car digital AI assistant. In Proceedings of the 11th International Conference on Automotive User Interfaces and Interactive Vehicular Applications: Adjunct Proceedings, Utrecht, The Netherlands, 21–25 September 2019; pp. 450–455.
104. Korthauer, A.; Guenther, C.; Hinrichs, A.; Ren, W.; Yang, Y. Watch Your Vehicle Driving at the City: Interior HMI with Augmented Reality for Automated Driving. In Proceedings of the 22nd International Conference on Human-Computer Interaction with Mobile Devices and Services, Oldenburg, Germany, 5–8 October 2020; pp. 1–5.
105. Liu, H.; Taniguchi, T.; Tanaka, Y.; Takenaka, K.; Bando, T. Visualization of Driving Behavior Based on Hidden Feature Extraction by Using Deep Learning. *IEEE Trans. Intell. Transp. Syst.* **2017**, *18*, 2477–2489. [CrossRef]
106. Dahl, D.A. *Multimodal Interaction with W3C Standards*; Springer International Publishing: Cham, Switzerland, 2017.
107. Pesek, M.; Strle, G.; Kavčič, A.; Marolt, M. The Moodo dataset: Integrating user context with emotional and color perception of music for affective music information retrieval. *J. New Music. Res.* **2017**, *46*, 246–260. [CrossRef]
108. Zhang, J.; Wang, B.; Zhang, C.; Xiao, Y.; Wang, M.Y. An EEG/EMG/EOG-Based Multimodal Human-Machine Interface to Real-Time Control of a Soft Robot Hand. *Front. Neurorobotics* **2019**, *13*, 7. [CrossRef]
109. Djamal, E.C.; Fadhilah, H.; Najmurrokhman, A.; Wulandari, A.; Renaldi, F. Emotion brain-computer interface using wavelet and recurrent neural networks. *Int. J. Adv. Intell. Inform.* **2020**, *6*, 1–12. [CrossRef]
110. Nam, C.S.; Nijholt, A.; Lotte, F. *Brain–Computer Interfaces Handbook: Technological and Theoretical Advances*; CRC Press: Boca Raton, FL, USA, 2018.
111. Ceccacci, S.; Mengoni, M.; Andrea, G.; Giraldi, L.; Carbonara, G.; Castellano, A.; Montanari, R. A Preliminary Investigation Towards the Application of Facial Expression Analysis to Enable an Emotion-Aware Car Interface. In *International Conference on Human-Computer Interaction*; Springer: Cham, Switzerland, 2020; pp. 504–517.
112. Delbouys, R.; Hennequin, R.; Piccoli, F.; Royo-Letelier, J.; Moussallam, M. Music mood detection based on audio and lyrics with deep neural net. *arXiv* **2018**, preprint. arXiv:1809.07276.
113. Ünal, A.B.; de Waard, D.; Epstude, K.; Steg, L. Driving with music: Effects on arousal and performance. *Transp. Res. Part F Traffic Psychol. Behav.* **2013**, *21*, 52–65. [CrossRef]
114. Amini, R.; Willemsen, M.C.; Graus, M.P. Affective Music Recommender System (MRS): Investigating the Effectiveness and User Satisfaction of different Mood Inducement Strategies. 2019. Available online: https://pure.tue.nl/ws/portalfiles/portal/131839906/Affective_MRS_R._Amini_v1.1.pdf (accessed on 31 June 2021).
115. Park, M.; Thom, J.; Mennicken, S.; Cramer, H.; Macy, M. Global music streaming data reveal diurnal and seasonal patterns of affective preference. *Nat. Hum. Behav.* **2019**, *3*, 230–236. [CrossRef] [PubMed]
116. Febriandirza, A.; Chaozhong, W.; Zhong, M.; Hu, Z.; Zhang, H. The Effect of Natural Sounds and Music on Driving Performance and Physiological. *Eng. Lett.* **2017**, *25*, 455–463.
117. Navarro, J.; Osiurak, F.; Gaujoux, V.; Ouimet, M.C.; Reynaud, E. Driving under the influence: How music listening affects driving behaviors. *J. Vis. Exp.* **2019**, *145*, e58342. [CrossRef]
118. Green, P. Crashes Induced by Driver Information Systems and What Can Be Done to Reduce Them (No. 2000-01-C008). Available online: https://citeseerx.ist.psu.edu/viewdoc/download?doi=10.1.1.589.3723&rep=rep1&type=pdf (accessed on 31 June 2021).
119. Li, B.; Sano, A. Extraction and Interpretation of Deep Autoencoder-based Temporal Features from Wearables for Forecasting Personalized Mood, Health, and Stress. *Proc. ACM Interact. Mob. Wearable Ubiquitous Technol.* **2020**, *4*, 1–26. [CrossRef]
120. Mikoski, P.; Zlupko, G.; Owens, D.A. Drivers' assessments of the risks of distraction, poor visibility at night, and safety-related behaviors of themselves and other drivers. *Transp. Res. Part F Traffic Psychol. Behav.* **2019**, *62*, 416–434. [CrossRef]
121. Bran, E.; Bautu, E.; Popovici, D.M. Open Affordable Mixed Reality: A Manifesto. In Proceedings of the 2020 International Conference on Development and Application Systems (DAS), Suceava, Romania, 21–23 May 2020; pp. 177–184.
122. Augusto, J.C.; Callaghan, V.; Cook, D.; Kameas, A.; Satoh, I. Intelligent environments: A manifesto. *Hum. Comput. Inf. Sci.* **2013**, *3*, 1–18. [CrossRef]
123. Schipor, O.-A.; Vatavu, R.-D.; Vanderdonckt, J. Euphoria: A Scalable, event-driven architecture for designing interactions across heterogeneous devices in smart environments. *Inf. Softw. Technol.* **2019**, *109*, 43–59. [CrossRef]
124. Schipor, O.A.; Vatavu, R.-D. Empirical Results for High-definition Video and Augmented Reality Content De-livery in Hyper-connected Cars. *Interact. Comput.* **2021**, *33*, 3–16. [CrossRef]
125. Bran, E.; Sburlan, D.F.; Popovici, D.M.; Puchianu, C.M.; Bautu, E. In-vehicle Visualization of Data by means of Augmented Reality. *Procedia Comput. Sci.* **2020**, *176*, 1487–1496. [CrossRef]
126. Sburlan, D.F.; Bautu, E.; Puchianu, C.M.; Popovici, D.M. Adaptive Interactive Displaying System for In-Vehicle Use. *Procedia Comput. Sci.* **2020**, *176*, 195–204. [CrossRef]

127. Bran, E.; Bautu, E.; Popovici, D.M.; Braga, V.; Cojuhari, I. Cultural Heritage Interactive Dissemination through Natural Interaction. In Proceedings of the International Conference on Human-Computer Interaction RoCHI, Bucharest, Romania, 17–18 October 2019; pp. 156–161.
128. Zaiți, I.-A.; Pentiuc, G.; Vatavu, R.-D. On free-hand TV control: Experimental results on user-elicited gestures with Leap Motion. *Pers. Ubiquitous Comput.* **2015**, *19*, 821–838. [CrossRef]
129. Shao, L. Hand Movement and Gesture Recognition using Leap Motion Controller; Virtual Reality, Course Report. Available online: https://stanford.edu/class/ee267/Spring2016/report_lin.pdf (accessed on 31 June 2021).
130. Bautu, E.; Tudose, C.I.; Puchianu, C.M. In-Vehicle System for Adaptive Filtering of Notifications. In Proceedings of the International Conference on Human-Computer Interaction RoCHI, Bucharest, Romania, 17–18 October 2019; pp. 145–151.
131. Bautu, E.; Puchianu, C.M.; Bran, E.; Sburlan, D.F.; Popovici, D.M. In-Vehicle Software System for Fostering Driver's Attentiveness. In Proceedings of the 2020 International Conference on Development and Application Systems (DAS), Suceava, Romania, 21–23 May 2020; pp. 151–156.
132. Naujoks, F.; Wiedemann, K.; Schömig, N.; Hergeth, S.; Keinath, A. Towards guidelines and verification methods for au-tomated vehicle HMIs. *Transp. Res. Part F Traffic Psychol. Behav.* **2019**, *60*, 121–136. [CrossRef]
133. Tan, W.-S.; Liu, D.; Bishu, R. Web evaluation: Heuristic evaluation vs. user testing. *Int. J. Ind. Ergon.* **2009**, *39*, 621–627. [CrossRef]

Article

User Evaluation of a Multi-Platform Digital Storytelling Concept for Cultural Heritage

Silviu Vert [1,*], Diana Andone [2], Andrei Ternauciuc [1], Vlad Mihaescu [1], Oana Rotaru [1], Muguras Mocofan [1], Ciprian Orhei [1] and Radu Vasiu [1]

1. Communications Department, Politehnica University of Timisoara, 300223 Timisoara, Romania; andrei.ternauciuc@upt.ro (A.T.); vlad.mihaescu@upt.ro (V.M.); oana.rotaru@student.upt.ro (O.R.); muguras.mocofan@upt.ro (M.M.); ciprian.orhei@cm.upt.ro (C.O.); radu.vasiu@upt.ro (R.V.)
2. eLearning Center, Politehnica University of Timisoara, 300223 Timisoara, Romania; diana.andone@upt.ro
* Correspondence: silviu.vert@upt.ro

Abstract: Digital storytelling platforms have proven to be a great way of bringing cultural heritage closer to people. What lacks is a deeper understanding of the user experience of such systems, especially in multi-platform digital storytelling. For the last three years, we have been developing a project called Spotlight Heritage Timisoara, which is at its core a digital storytelling platform for the city of Timisoara (Romania), soon to be European Capital of Culture in 2023. The project consists of a website, mobile applications, and interactive museographic and street exhibitions. This paper presents a multi-platform usability evaluation study which employed semi-structured interviews, observations, think-aloud protocol, SUS questionnaire, Net Promoter Score and Product Reaction Cards to gather insights from 105 participants and reveal usability problems in the Spotlight Heritage context. We found out that the four platforms, i.e., interactive touchscreen table, desktop/laptop, mobile and Augmented Reality, have very good usability scores, are considered accessible and useful, work seamlessly together, and create user satisfaction and loyalty, across demographic groups, having the potential to bring people closer to cultural heritage.

Keywords: digital storytelling; cultural heritage; usability evaluation; multi-platform evaluation

1. Introduction

Digital Storytelling is a successful combination between the ancient and time-proven art of telling a story and the possibilities of technology today. It has been applied in various domains, one of the most interesting areas being the Cultural Heritage, where it has proven its capabilities of bringing heritage elements closer to people.

Cultural Heritage is experimented in many forms today, from visiting a museographic exhibition to surfing the museum's website or mobile app or using Augmented Reality applications while in front of artworks. Because of this, there is a strong need for testing and adapting the user experience of such digital products in order to provide a fulfilling experience to the visitor.

In the next subsections, we introduce a multi-platform digital storytelling concept that we developed and tested. We also describe the research aims of this paper.

1.1. Project Description

Spotlight Heritage Timisoara is a digital cultural initiative of the Politehnica University of Timisoara, through the eLearning Centre and the Multimedia Center, realized in partnership with the National Museum of Banat, part of the Timisoara European Capital of Culture program. Spotlight Heritage Timisoara reveals, by digital storytelling, the city of Timisoara through stories of cultural and historical heritage, technical development, communities, and neighborhoods, interwoven with the personal stories of the inhabitants of yesterday and today.

The open, participatory mobile applications and web platform contain Landmarks (data and information of cultural and historical heritage, geographically mapped, photos and films, with recognition of buildings by augmented reality—with information and old pictures superimposed, as well as 360 images), Sequences (personal stories of the inhabitants which provide a historical, social, and cultural background to the landmarks in the neighborhood presented in the classic exhibition in the attic of the Maria Theresia Bastion), Community (information from the history of neighborhoods, communities, ethnic groups, organizations that were or are present in the neighborhood), Your story (users can add their own personal stories, comments, photos, videos), and Events (events from exhibitions, guided tours offered with high school teenagers).

Spotlight Heritage Timisoara was planned to be publicly delivered as a multi-platform environment—on an interactive display in the National Banat Museum, on a dedicated website, as a mobile app, and as an augmented reality mobile app (Figure 1). The information provided with each of these delivery devices is similar, in some cases identical, but the displayed information differs, as dictated by their technical affordances, hence each creating a different user experience.

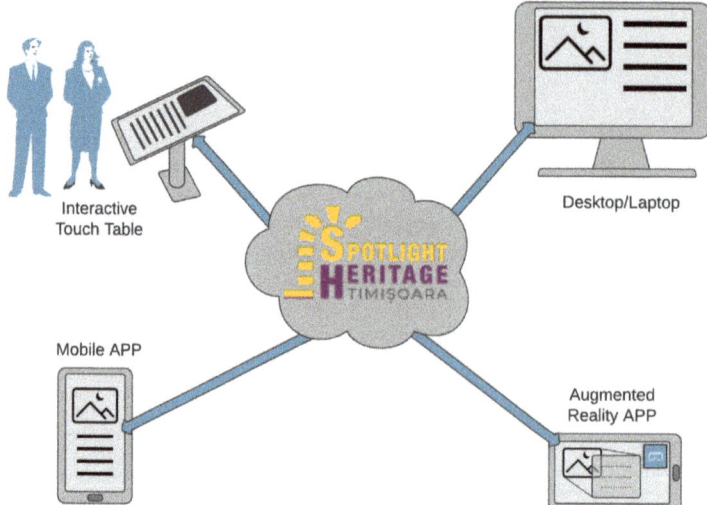

Figure 1. Multi-platform architecture of Spotlight Heritage Timisoara.

1.2. Spotlight Heritage Evaluation Study

These complex digital products have been evaluated in a multi-platform setting, i.e., touchscreen table, desktop/laptop, mobile, and Augmented Reality. The user evaluation followed a mixed methods approach, to respond to the stringent need for multiple platform evaluation and comprehensive user testing [1]. Spotlight Heritage is a multi-annual project and from our previous evaluation and use cases, we identified that users navigate seamlessly between different platforms—they might initiate it online, on the website, then move and continue the exploration on the mobile app, go and visit the physical exhibition and use the interactive display, or visit one of the landmarks in location and use the augmented reality app.

Our research paper aims to indicate our research results so as to:

(1) Validate the technical design and implementation of the multi-platform digital storytelling concept for the neighborhoods of Timisoara, European Capital of Culture in 2023.

(2) Identify the user evaluation, methodology, and process of the multi-platform digital products developed as part of the cultural concept.

(3) Provide insights and good practice on how to improve the user experience of such a multi-platform digital storytelling project.

First of all, qualitative and quantitative data were collected via a survey that was sent to participants of the opening of the Elisabetin exhibition, in 2020. The survey, which was completed by 75 respondents, included demographic questions, System Usability Scale questionnaires and Net Promoter Score (NPS) questions. Secondly, the digital applications have been tested with 30 participants using a combination of in-person and remote moderated usability testing, System Usability Scale questionnaires, and Product Reaction Cards.

1.3. Organization of the Paper

In Section 2, we discuss best practices and challenges in designing and implementing digital storytelling platforms in Cultural Heritage. In Section 3, we describe the planning, architecture, and implementation of the Spotlight Heritage project. In Section 4, we describe the methodology used to evaluate the digital storytelling platform, showing how it is grounded in similar peer-reviewed research. In Section 5, we present the results of the evaluation and some discussions. Section 6 is dedicated to an overall discussion of the results of this research, while Section 7 ends the paper with general conclusions.

2. Background

2.1. Digital Storytelling Platforms in Cultural Heritage

Digital Storytelling is short, narrated films (2 to 5 min) best known because of the work of the Centre of Digital Storytelling (CDS) in Berkeley, California. Originating in the arts community, Digital Storytelling has been adapted to work very well in healthcare, business, education, research, or cultural heritage [2,3].

The CDS digital stories are personal reflections on a subject, are revealing insights of the author, are narrating a lived experience, are based on photos and soundtracks, are short in length (for brevity), and are about self-expression and self-awareness (over concerns of publication and audience) [4].

One of the most interesting applications for Digital Storytelling is in Cultural Heritage, where it demonstrated its effectiveness in communicating heritage interpretation to the public [5]. Cultural Heritage, in Bill Ivey's words, "tells us where we came from by preserving and presenting voices from the past, grounding us in the linkages of family, community, ethnicity, and nationality, giving us our creative vocabulary" [6].

Digital Storytelling engages the museum visitor by combining the powerful mechanism of the story with the possibilities of the digital technologies. Firstly, it is a vehicle for edutainment, meaning that it combines the instructive part of the museum with the entertaining part that the visitor expects. Secondly, Digital Storytelling enhances the visitor experience, transforming the static and traditional form of museum exhibitions into lively, accessible, and engaging experiences [7].

These types of platforms have been designed and implemented in several culturally rich places.

In ref. [5], the authors describe the design, implementation, and evaluation of a mobile digital storytelling experience for the Athens University History Museum. The evaluation focuses both on the technological aspects and on the story perspective. The evaluation confirmed that their approach to digital storytelling promoted engagement, learning and deeper reflection, while it also revealed several insights for good practice.

In ref. [8], the authors describe the development and evaluation of SensiMAR, a Multisensory Augmented Reality system for outdoor cultural heritage, implemented in the Conimbriga Ruins, part of the Monographic Museum of Conimbriga National Museum (Portugal). The prototype combines visual, audio and olfative contents.

In ref. [9], the authors focus on digital storytelling strategies that promote active audience engagement in urban cultural heritage. They propose a collaborative model, which fuels audience engagement and collaboration of cultural organizations, and build a prototype for the culturally rich city of Thessaloniki (Greece).

In ref. [10], the authors use Interactive Digital Storytelling to enhance the visitor experience at the White Bastion, which is a fortress overlooking the city of Sarajevo. Ten digital stories and 6 interactive virtual models of the fortress have been made available for the users.

Details on the evaluation of these platforms will be given later, in our Methodology section.

2.2. Usability Evaluations

Usability is considered to be one of the most important factors for the use of web or mobile applications, along with reliability or security. The challenge to develop more usable web or mobile applications has led to various methods, techniques, and tools in order to discover usability issues [11].

According to [12], even if it is performed as part to the design process or after the product was launched, the evaluation of a cultural heritage application plays a crucial role in its usability or acceptance.

Being a multi-platform app, the usability evaluation of Spotlight Heritage is really important when it comes to testing functionality on different devices, to enhance the general design on web or mobile, intuitiveness, efficiency, effectiveness, users' satisfaction [13].

A simple way of evaluating the usability of a platform is the System Usability Scale questionnaire (SUS) [14], which was also applied in our case.

SUS is a standardized questionnaire, with alternating questions, half of them worded positively and half negatively. According to [15], the advantages of positive and negative items in a survey can be the ability to provide protection against the extreme responders while also keeping the participants engaged, being careful for each question's tone. There are also disadvantages for alternating questions when it comes to the possibility of misinterpreting them when the users do not pay enough attention to the item's tone or miscoding when it comes to calculate the SUS score.

There are many other usability evaluation questionnaires like SUMI, PSSUQ, or QUIS [16], but after researching which one is best applicable in our case, we decided that SUS is the proper choice, being short, easy, and in an objective tone.

3. Development of the Spotlight Heritage Multi-Platform Digital Storytelling Concept

As previously stated, the Spotlight Heritage Timisoara project is an integral part of the application bid with which the city of Timisoara won the title of European Capital of Culture 2021 (transformed into European Capital of Culture 2023 due the global COVID-19 pandemic). It aims not only to provide information to visitors, in the form of presentations of objectives, photos, short videos, but is also designed to be interactive in the sense that it offers them the opportunity to post their own stories and memories of the neighborhoods through the website and the applications. This way, users become co-creators of the exhibition, through a participatory approach and responding to the objectives with which Timisoara won the title of European Capital of Culture: involvement and activation of various audiences. Visitors and citizens of Timisoara add their personal layer over that of the exhibition discourse and the historical reality, enriching and keeping the city's heritage alive.

3.1. Spotlight Heritage Planning

There are two main target groups of people that the project aims to reach. Firstly, the visitors who will be present, mostly physically but virtually as well, in the city of Timisoara around the year 2023. They will learn about the history of the city through these visits, which means that they need to be memorable. Secondly, the citizens currently residing here, many of whom can trace their ancestral roots hundreds of years into the city's history. They need to be able to reconnect with their forebears and cement their sense of belonging. All these people need to be able to leave their mark into the tapestry of Timisoara's continuing story.

Planning and designing the implementation of the project considered the stated objectives: to present information in an engagingly interactive fashion and to encourage the audience participation in the cultural act by transforming passive consumption of experiences into active contribution.

Taking all these considerations into account, the project was designed with multiple distinct components in mind:

- The physical yearly expositions hosted by the National Museum of Banat;
- Location-based expositions, highlighting different historically significant neighborhoods from Timisoara;
- The digital components.

Even if these components seem unrelated through their scope, target audience and implementation methods, they complement each other, and by a carefully orchestrated design, manage to bring to life Timisoara's rich history.

In order to increase interactivity, classical exhibitions are being enhanced through modern technologies. Visitors in the museum or in the neighborhoods can access more information than is available on site through the use of digital tools which they either own or are provided on location.

The exhibitions hosted by the National Museum of Banat, for instance, provide stand-mounted tablets and an interactive touch table which they can use to access additional information regarding past or present exhibits. The large table promotes cooperation and discussion as well as an unusual and exciting medium which provides related content, while the tablets give a taste of advanced interactions (like augmented reality) which the visitors may be enticed to replicate by themselves later.

The stations placed in relevant point-of-interest alongside the current spotlighted neighborhood provide printed QR codes which can be accessed by appropriate applications on the visitors' own mobile devices for further content, increasing the sense of active participation.

These complex scenarios involving physical visits by different types of participants, combined with virtual users who can access many of the same resources from the Internet, have provided the inspiration for the multi-platform architecture that forms the basis of the Spotlight Heritage Timisoara Project.

3.2. Spotlight Heritage Multi-Platform Architecture

Developing the software architecture was heavily dependent on the multi-platform architecture (Figure 1) based on the hardware being used in each of the envisioned scenarios. The display terminals provided on location represented known quantities: the tablets have specific versions of the Operating Systems and technical specifications, and the interactive touch table can support—in the simplest use case scenario—a web browser with online access. However, the personal devices that visitors use in the museum or in the neighborhoods can vary widely in specifications, capabilities, and performance. The same goes for the terminals that virtual visitors use to access via the Internet.

The most versatile medium of providing the virtual experience was therefore the website. A modern, adaptive, and responsive platform was developed which can be accessed from most Internet capable terminals. This platform provides a classic navigation experience for accessing the virtual exhibits which are usually static images with extended text descriptions, as well as audio and video content. The exhibits are grouped by collections and can be filtered or visited in a guided tour, if desired.

A particular use case of the website is its access on the large interactive touch table, by multiple visitors simultaneously, and in a crowded museum setting. This innovative use case is of particular interest and has separate sections of this paper dedicated to it.

While the website can be efficiently accessed on a mobile device with a smaller screen, current web technologies cannot fully take advantage of all the facilities that a mobile smartphone or tablet has to offer. Accessing sensors such as the gyroscope (for orientation), the location, or even the camera through the mobile browser is not fully optimized, or problematic at times due to privacy issues [17]. In order to access the greatest number of users possible, native mobile applications were developed for the two main mobile ecosystems: Apple's iOS and Google's Android Operating Systems (OS). These applications benefit from high compatibility and integration with the host OS as well as native access to location, providing a seamless experience with which users are accustomed. The application provides the same virtual exhibits as the website, as well as dynamic positioning and real-time recommendations for nearby physical exhibits.

A more advanced use case of the mobile device is through the Augmented Reality capabilities that some terminals can provide. Using image recognition, it is possible to superimpose additional information to the images captured in real-time by a device's camera. In order to keep the compatibility with as many devices as possible, it was decided that different applications needed to be developed solely for this functionality. Therefore, the Augmented Reality Applications (for iOS and Android) were created.

These are the main channels through which the virtual content encompassed in the Spotlight Heritage Timisoara project can be accessed. Figure 2 aims to synthesize in a graphical form the architecture of the digital components of the project and highlights the four directions for evaluation that this research paper analyses.

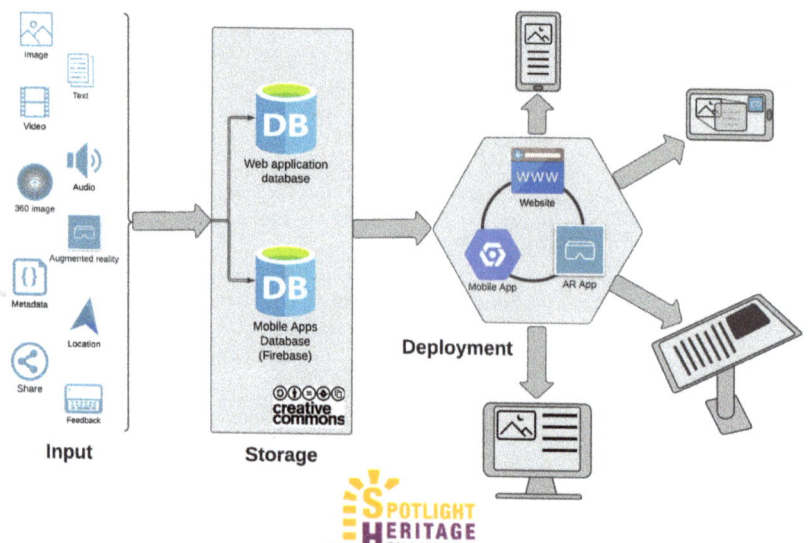

Figure 2. Data architecture and flow.

3.3. Spotlight Heritage Implementation

Following the developed architecture, the different parts of the project were subsequently implemented.

The physical components (the temporary exhibitions at the National Museum of Banat and on location in each of the neighborhoods) were staged with a different focus in each of the previous years. In 2019, the focus was on the Iosefin neighborhood, in 2020 on the Elisabetin neighborhood, and in 2021 on the Fabric neighborhood.

The temporary exhibits at the Museum featured objects representative for the respective neighborhood, with the added interactivity provided by the large interactive touch table and stands with tablets showcasing the mobile applications (Figure 3).

Figure 3. Large Interactive Touch Table at the National Museum of Banat exhibition (2019).

During each of the three years since the beginning of the project, near the landmarks of the highlighted neighborhood, informational stations were positioned with descriptions of the landmark, a short history, and the invitation of using the available technologies (website or mobile applications) to find further details about that specific landmark, and other similar ones. Stations were created and deployed for 18 locations in the Iosefin neighborhood, 16 in Elisabetin and 23 in Fabric (Figure 4).

Figure 4. Street stand with project information near a neighborhood landmark (2019).

Both the museum exhibitions and the stations displayed on location were active only temporarily. However, all the information was preserved digitally in the third component of the architecture described in the Section 3.1.

The main repository of information is available through the online web platform, at https://spotlight-timisoara.eu (accessed on 10 October 2021). Pictures with descriptions and other related metadata are available, categorized into the three main neighborhoods (Figure 5), as well as audio and video excerpts, related to the different landmarks from each of the locations. The platform is optimized for access from any size screen from the very small (mobile terminal) to the very large (interactive touch table).

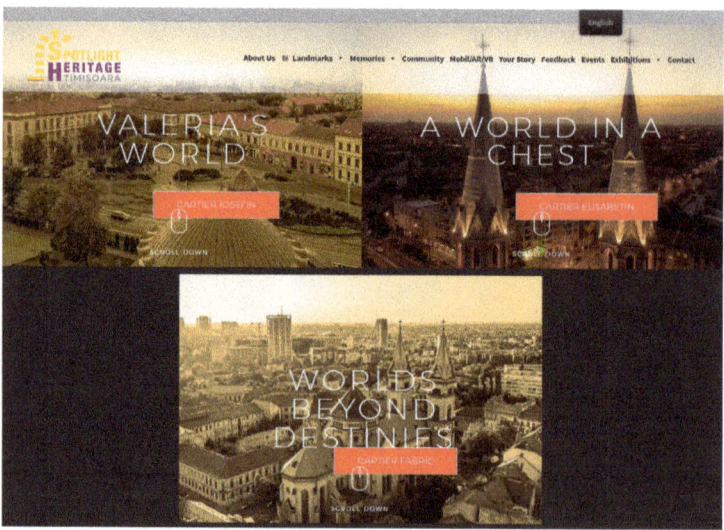

Figure 5. The Spotlight Heritage Timisoara website in 2021.

Development also included a total of four mobile applications: the mobile and augmented reality apps, each with a different version for the two major mobile operating systems: iOS and Android.

The mobile application provides most of the same information as the web platform, as well as easy navigation towards the landmarks showcased in the project (Figure 6).

The augmented reality app offers advanced facilities, such as superimposing old photos over the current landmark, which is being filmed in real-time, or by adding information cards over landmarks which it recognizes (Figure 7).

Both applications are available for free in the respective application stores, Apple's App Store and Android's Google Play Store.

As a rough estimate, the human effort that was put in during the lifetime of the project so far is 120 persons/month, which includes the 2014–2018 period in which the concept of the project emerged and was refined and the first 3 years of implementation (2019–2021). The first year of implementation demanded the most effort, around 50 persons/month, since most of the physical and software infrastructure was created at this point.

Each cycle of implementation, i.e., adding new information and developing additional features, involved a plethora of competencies and professions: architects, historians, anthropologists, curators, web developers, mobile developers, data specialists, (video) photographers, audio-video editing specialists, augmented reality professionals, event managers, public relation specialists, and many more.

A lesson learned is that such a complex multi-platform digital storytelling project for cultural heritage requires a dozen of professionals working interdisciplinary in a smooth and efficient manner, time and budget accordingly, and high attention to how the user experience integrates all the components.

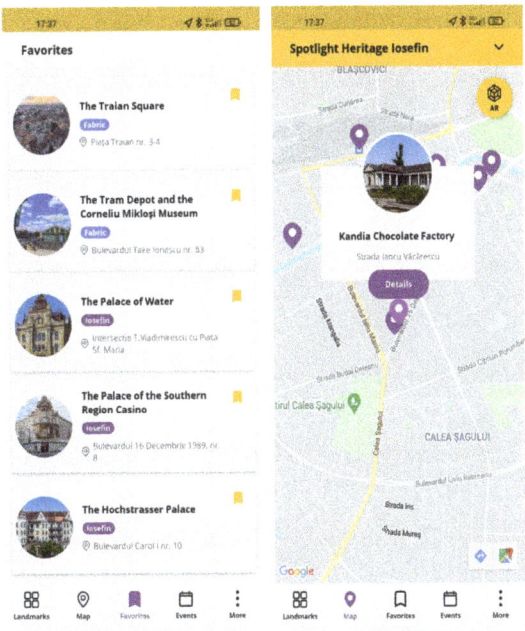

Figure 6. Screen captures from the Spotlight Heritage Timisoara mobile application in 2021.

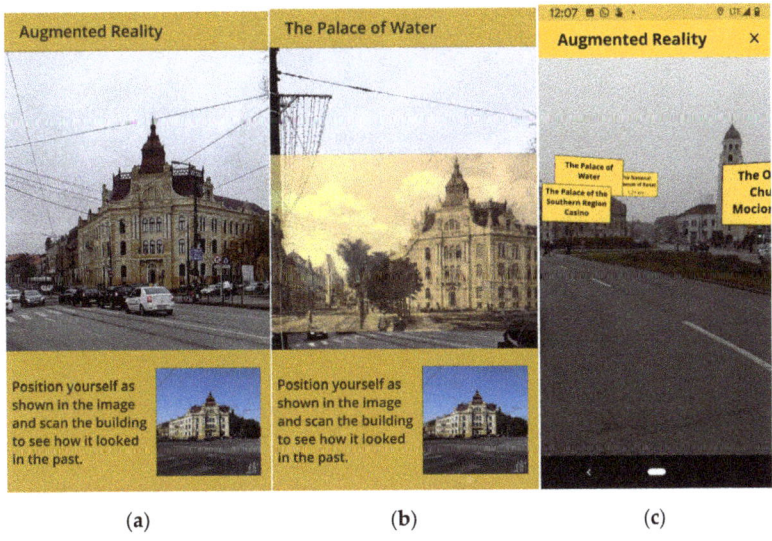

Figure 7. Features of the Augmented Reality App: (**a**) original image, captured real-time in the application, (**b**) historical image, over-imposed on the screen, (**c**) landmarks which are identified and labeled by the app.

4. Spotlight Heritage Evaluation Methodology

4.1. Related Work

Our evaluation methodology is grounded in similar peer-reviewed research.

The app for Athens University History Museum in [5] was evaluated with 43 museum visitors, of which 28 invited participants and 15 participants recruited on the spot. The authors used a combination of semi-structured interviews, focus group sessions, and user observation sessions.

The authors of SensiMART [8] performed a usability study on 67 random visitors of the archaeological site. The participants were invited to explore the application (with minimal intervention from the researchers) and then fill a SUS questionnaire followed by a generic socio-demographic questionnaire. Researchers found that there are no statistically significant differences in the usability of the product, when considering sex, age or previous experience with the technology.

The researchers in Thessaloniki [9] performed a formative qualitative and quantitative evaluation prior and after the development of the model. Specifically for the usability of the pilot website, a quantitative assessment was conducted, based on the heuristic analysis metrics by Jakob Nielsen.

4.2. Methods

In the following subsections, we briefly describe the evaluation methods which we have used throughout our study.

4.2.1. Semi-Structured Interviews

The Semi-structured Interview is a blend between open- and closed-ended questions and it is also a method often used in the usability evaluation of a web or mobile app [18].

4.2.2. User Observation Sessions

The User Observation session also known simply as User Testing is one of the most important and used usability evaluation methods and it is clearly one of the most applicable methods for the evaluation of a web or mobile product.

As explained in [19], User Testing evaluations are usability evaluation methods where the users actively participate. They are invited to do some typical tasks for a product or simply asked to explore the application/website to observe design issues or functional errors of the product. During these sessions, the time taken to do a task, the task completion rates, and the number and types of errors are recorded [4]. This method usually takes place in a designated laboratory, properly equipped with cameras which record the activities. Additionally, a group of observers, a moderator and a facilitator is involved in the user testing session [4].

4.2.3. Think-Aloud Protocol

The Think-aloud protocol is a method used also in the usability evaluations and it implies that the participants are verbalizing their thoughts while doing the requested task. The researchers record the comments and analyze them afterwards in a way that depends on the research questions [20].

4.2.4. System Usability Scale (SUS) Questionnaire

According to [14], SUS is an instrument which is used in the usability testing of commercial products. However, it can be a useful way of evaluating websites, desktop/mobile applications, which is also our case. Basically, SUS is an easy-to-use questionnaire consisting of 10 simple statements. The tool measures the users' perception of agreement or disagreement with 10 aspects of the product under evaluation [21]. The questions use mixed scaling and the final score is a number between 0 and 100.

An adjective rating scale ("Worst Imaginable", "Awful", "Poor", "OK", "Good", "Excellent", "Best Imaginable") can be used to explain the final score [22].

4.2.5. Net Promoter Score

According to [23], the Net Promoter Score is linked to the measurements of loyalty of the target group of a product and it is a very useful instrument for the managers, but also for the researchers, which is our case.

4.2.6. Product Reaction Cards

The Product Reaction Cards method is a tried and tested evaluation tool in the software development environment. It can be realized in a physical way, with physical cards to be picked by the participants, or by electronic means with the help of an Excel or Word file. According to [24], the Product Reactions Cards were initially created by Microsoft as part of a "desirability toolkit" which they used to measure the level of desirability in using a certain software product. The initial kit consisted of two components: the first one—a "study of the faces", where participants were tasked with choosing a picture of a face seeming to express their own experience with using the product, and the second component—a card study, where participants had to choose a few appropriate words from a large, predefined set of product reaction cards [5].

4.2.7. Error Testing

The Error testing method is somehow included in the usability testing method, being targeted on the free exploration of the application done by the participants, in order to find functional errors of the product.

4.3. Participants

The following subsection describes the two main stages of the user evaluation, i.e., an online survey and a set of usability testing sessions, and the participant demographic profiles.

4.3.1. Survey Respondents

An online survey with 75 respondents was conducted in November 2020, during the opening of the exhibition for a new neighborhood (Elisabetin) in Spotlight Heritage Timisoara. The survey included demographic questions (Table 1), separate SUS and NPS questionnaires for the desktop and the mobile versions of the digital storytelling concept and other questions regarding the usability and usefulness of the digital applications and cultural heritage data.

4.3.2. Usability Evaluation Participants

A usability evaluation with 25 participants (Table 2) was conducted in May 2021 using multiple methods: semi-structured interviews, observation sessions, think-aloud protocol, SUS questionnaire, and Product Reaction Cards. The testing sessions were organized both in-person and remotely by students enrolled in the Interactivity and Usability graduate class at the Politehnica University of Timisoara, under the supervision of the authors of this research, using a pedagogical approach already presented in [25] for website usability testing and in [26] for mobile usability testing. They recruited 10 participants for testing the desktop/laptop version, 10 participants for the mobile version, and 5 participants for the augmented reality version.

Another smaller usability evaluation was conducted directly by the researchers in July 2021, during the opening of the exhibition for the newest neighborhood (Fabric) in Spotlight Heritage. No demographic data were gathered for these additional 5 participants.

Table 1. Demographic profile of the survey respondents.

Criteria	Variation	Number	Percent (%)
Gender	Male	59	78.66
	Female	16	21.33
Age	<18	0	0.00
	18–26	41	54.67
	26–35	3	4.00
	35–50	18	24.00
	50–65	9	12.00
	>65	4	5.33
Professional background	Teacher	29	38.67
	Student	41	54.67
	Public administration	3	4.00
	Other	2	2.67
Residency	Timisoara	38	50.66
	Other cities from Romania	37	48.00
	Other country	1	1.34
Discover criteria of event	Media platform (traditional or social)	24	26.09
	Other people	21	22.83
	Other sources	47	51.09
Familiarity with Spotlight Heritage	First time	49	65.33
	Used it in the past	26	34.67

Table 2. Demographic profile of the usability evaluation participants.

Criteria	Variation	Number	Percent (%)
Gender	Male	8	38.10
	Female	13	61.90
Age	<18	0	0.00
	19–25	17	80.95
	26–35	4	19.05
	36–45	0	0.00
	46–55	0	0.00
	55–65	0	0.00
	>65	0	0.00
Professional background	Student	11	52.38
	Employee	9	42.86
	Self-employed	1	4.76
	Unemployed	0	0.00
Domain of activity	Art/culture	1	4.76
	IT	10	47.62
	Education	6	28.57
	Other	4	19.05
Educational level	Highschool	12	57.14
	Bachelor	4	19.05
	Master	5	23.81
	PhD	0	0.00
PC usage frequency	Once a week or less	3	14.29
	Once a day	1	4.76
	Several times a day	17	80.95
Smartphone usage level	Do not own a smartphone	0	0.00
	Beginner	0	0.00
	Average user	7	33.33
	Advanced user	14	66.67

4.4. Procedure

The following subsection describes the user evaluation procedure for each of the targeted platforms. The evaluation was then centralized to derive usability problems and insights, as indicated in Figure 8.

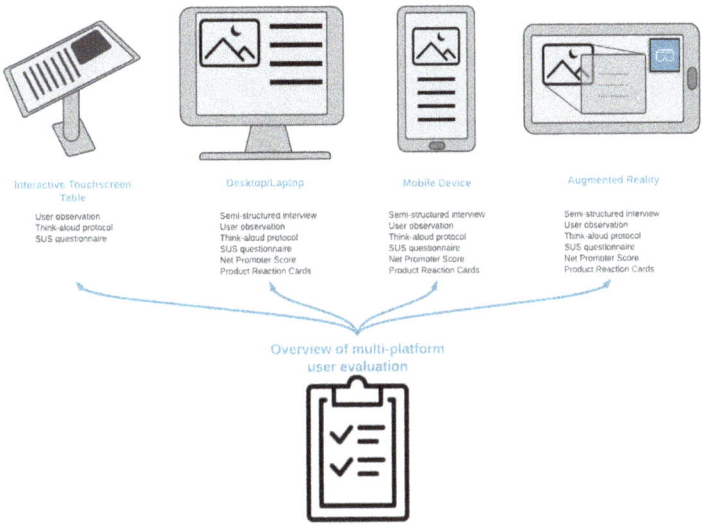

Figure 8. Mixed method approach for the multi-platform evaluation.

4.4.1. Evaluation Procedure for the Interactive Touchscreen Table

An informal in-person usability testing session was organized for the Spotlight Heritage platform running on the interactive touchscreen table (Figure 9). We randomly selected 5 participants who were trying out the interactive touchscreen table at the opening of the exhibition. We ran observation sessions with the think-aloud protocol and requested the participants to fill in a SUS questionnaire.

Figure 9. Users trying out the interactive touchscreen table platform, July 2021.

4.4.2. Evaluation Procedure for Desktop/Laptop

The user evaluation for desktop/laptop was organized remotely on Zoom, due to COVID-19 pandemic restrictions, in May 2021. Each session took approximately 60 min, and the evaluation was done with, in total, 10 participants.

Firstly, the participants signed a consent form which allowed us to record the meeting and process the resulting data for research purposes. Afterwards, they were instructed on how to share their browser in Zoom, and other technical aspects.

Secondly, we ran a semi-structured interview with the participants, whose aim was to find out their previous frustrations and satisfactions in using web platforms on desktop/laptop, their existing knowledge on the cultural heritage of Timisoara, and what kind of information would they be searching for (and how) on a website about this topic.

Thirdly, we ran an observation session with each participant. They were instructed to complete a list of pre-made tasks, consisting in exploring the homepage of the website, finding information about a landmark, adding a personal story, and participating in the online virtual tours. We were able to watch, measure and record what the participants did in the browser, as well as their face expressions. Using the think-aloud protocol, we encouraged the participants to speak what was on their mind while completing the task.

Fourthly, we verbally requested some short feedback about their interaction with the website and we handed them an online questionnaire where they were requested to answer the 10 SUS questions and to choose 5 words that best represented their experience with the website from the Product Reaction Cards set.

Finally, we thanked the participants for their time and noted down if they are interested in being notified about future events related to Spotlight Heritage Timisoara.

In Figure 10, we present a screenshot from the usability evaluation of the Spotlight Heritage website.

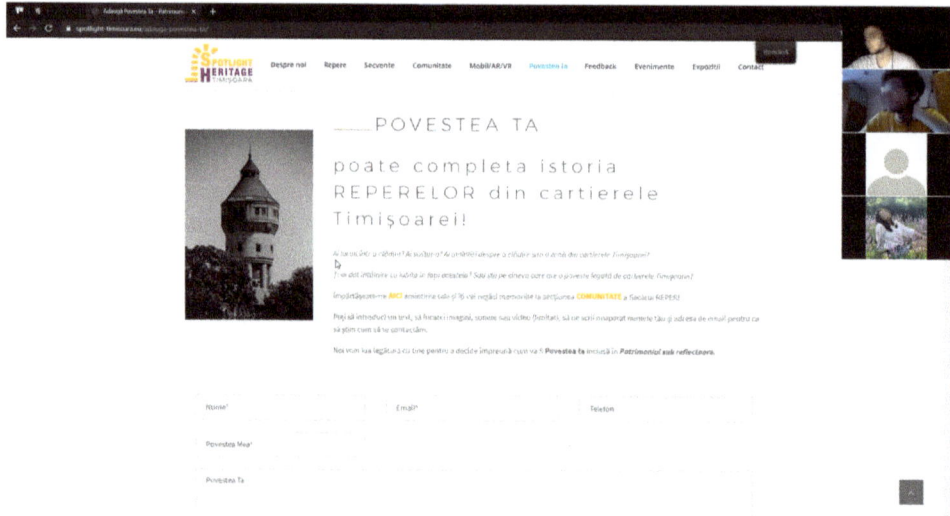

Figure 10. Spotlight Heritage website evaluation.

The desktop/laptop platform was also evaluated via a SUS questionnaire, an NPS question, and some open-ended questions through the 75-respondents online survey that was mentioned in the previous sections.

4.4.3. Evaluation Procedure for the Mobile Device

The evaluation on the mobile platform was conducted very similarly to the one on the desktop/laptop platform and with the same number of users (10), also in May 2021. What we noticed was a slightly higher technical difficulty for the participants, since they needed to join the meeting both from their computer, in order for us to follow what they say and their non-verbal feedback, and their smartphone, in order for them to be able to share with us how they complete the tasks in the mobile application (Figure 11).

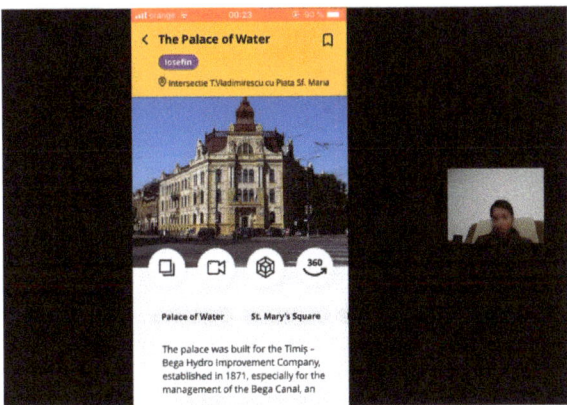

Figure 11. Usability evaluation for the mobile app, using remote conferencing tools.

Similarly to the desktop/laptop platform, feedback for the mobile version was collected also through the 75-respondents survey.

4.4.4. Evaluation Procedure for the Augmented Reality Application

The user evaluation for augmented reality was again performed very similarly to the evaluations described above. This time, however, it was an in-person moderated usability testing session, as augmented reality applications cannot be tested otherwise. Due to the pandemic, as it was performed in May 2021, we restricted the number of participants to 5, who met the moderators in front of the landmarks where the AR application was working (Figure 12). The number of tasks was reduced, the main one being the scanning of the landmark with the AR app.

Figure 12. Usability evaluation with the AR application in front of a landmark.

5. Spotlight Heritage Evaluation Results

5.1. Evaluation Results on the Interactive Touchscreen Table

The usability evaluation for Spotlight Heritage on the interactive touchscreen table revealed some positive and negative aspects. Some of the positive aspects mentioned by the participants were related to the capability to see the interface more clearly and to have a closer view. However, one of the disadvantages of navigating on the app on a device of this size is the necessity for wider gestures in order to reach buttons, menus, keyboards.

The users also mentioned the better user experience, since accessing the information as a part of a group allowed them to discover facts, data and information more easily, and discussing with the near-by users improved the social experience. This will be further investigated as part of our future research, seeking the social usability of our multi-platform cultural product [27].

The average SUS score for the interactive touchscreen table was 90, corresponding to the "Best Imaginable" usability adjective.

5.2. Evaluation Results on Desktop/Laptop

From the Spotlight Heritage website evaluation session, the emphasized positive aspects are referring to the simple yet very attractive user interface design, the existence of the map which is very useful for finding the buildings presented on the platform, the navigation which is quite simple and intuitive, and the presence of useful icons. Moreover, the possibility of adding your own memories related to some landmarks was highly appreciated by participants, bringing a higher level of interaction between the user and the platform and also increasing credibility and confidence among users.

On the other hand, the usability issues found on the website referred to the overcrowded menu, the lack of a search bar, the similarity between the icons of some buttons, and the fact that the syntagm "My story" was not very intuitive for its purpose.

As part of the usability testing sessions, we ran the SUS questionnaire which resulted in an average score of 92, corresponding to the "Best Imaginable" usability adjective. In addition, we ran the Product Reaction Cards method, the first 5 words chosen by the participants being "Complex", "Accessible", "Understandable", "Efficient", and "Useful".

On the other hand, through the online survey that we conducted, we were able to segment the SUS scores based on respondent demographics (Table 3 and Figure 13).

Table 3. Average usability scores and adjectives, presented by sex, age, and past experience with the Spotlight Heritage concept, for the desktop/laptop platform (data collected from the online survey and represented as a table).

Participant Group	Sample	Average Usability Score	Usability Adjective
Total	44	82.20	Excellent
Female	30	81.03	Excellent
Male	14	82.20	Excellent
Age < 18	0	0.00	-
Age 18–26	28	79.90	Excellent
Age 26–35	1	95.00	Best Imaginable
Age 35–50	7	79.64	Excellent
Age 50–65	7	90.71	Best Imaginable
Age > 65	1	87.50	Excellent
Past experience SH	17	82.50	Excellent
First experience SH	27	82.02	Excellent

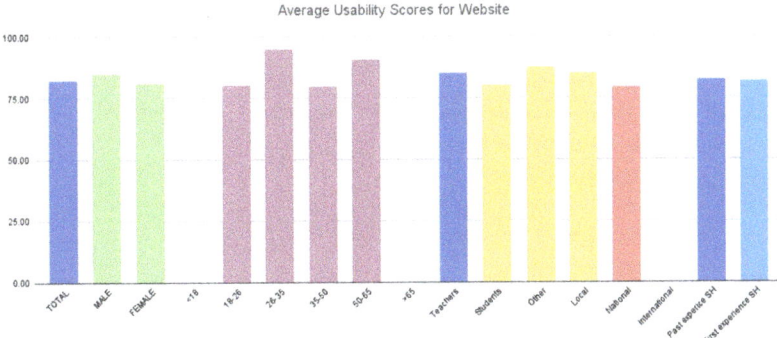

Figure 13. Average usability scores and adjectives, presented by sex, age, and past experience with the Spotlight Heritage concept, for the desktop/laptop platform (data collected from the online survey and represented as a graphic).

As part of the online survey, we also asked for feedback regarding specific aspects of the website (Figure 14).

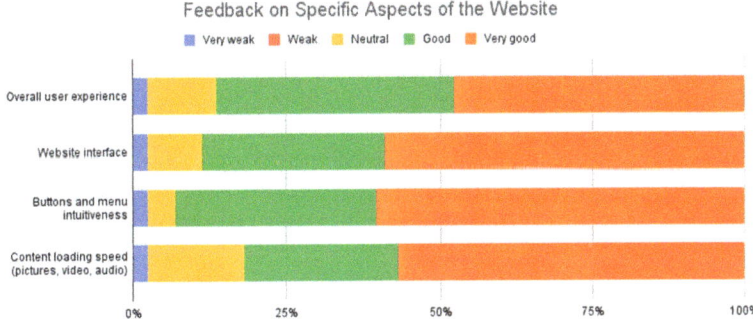

Figure 14. Respondent feedback regarding the user experience on the website (data collected from the online survey).

The average NPS score obtained by the desktop/laptop platform in the online survey was 68.09.

5.3. Evaluation Results on the Mobile Device

After the Spotlight Heritage mobile application evaluation session, the positive aspects found include the general easy and intuitive design, the matching colors of the app, the good functionality of the map, and the navigation which was considered quite easy.

The usability issues found here include the lack of a search bar, some non-intuitive icons and the long loading time for opening the app.

The average SUS score obtained after the usability testing sessions was 85.5, corresponding to the "Excellent" usability adjective. We applied again the Product Reaction Cards method; the first 5 words which were emphasized by the participants were "Captivating", "Accessible", "Attractive", "Useful", and "Good quality".

Similarly, through the online survey that we conducted, we were able to segment the SUS scores based on respondent demographics (Table 4 and Figure 15).

Table 4. Average usability scores and adjectives, presented by sex, age, and past experience with the Spotlight Heritage concept, for the mobile platform (data collected from the online survey and represented as a table).

Participant Group	Sample	Average Usability Score	Average SUS Quarter Grade
Total	43	79.19	Excellent
Female	32	77.97	Excellent
Male	11	82.73	Excellent
Age < 18	0	0.00	-
Age 18–26	33	78.56	Excellent
Age 26–35	2	96.25	Best Imaginable
Age 35–50	5	83.50	Excellent
Age 50–65	2	73.75	Good
Age > 65	1	55.00	OK
Past experience SH	17	79.12	Excellent
First experience SH	26	79.23	Excellent

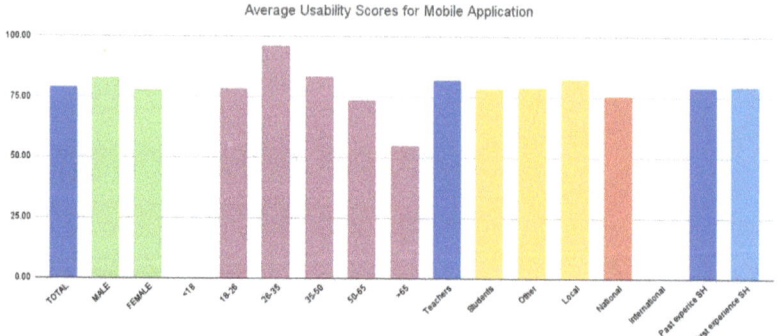

Figure 15. Average usability scores and adjectives, presented by sex, age, and past experience with the Spotlight Heritage concept, for the mobile platform (data collected from the online survey and represented as a graphic).

Similarly, as part of the online survey, we also asked for feedback regarding specific aspects of the mobile app (Figure 16).

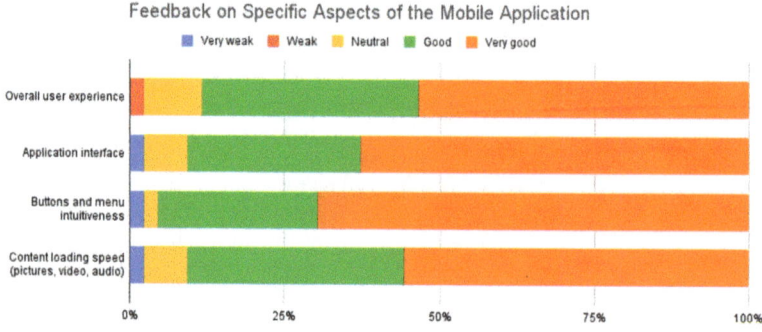

Figure 16. Respondent feedback regarding the user experience in the mobile app (data collected from the online survey).

The average NPS score obtained by the mobile platform in the online survey was 70.45.

5.4. Evaluation Results on the AR App

The evaluation of the Spotlight Heritage AR mobile app resulted in a general positive impression about the design and intuitiveness, as well as the easy flow of the navigation. The negative aspects discovered during the evaluation are based on the difficulties of positioning the smartphone camera in space in order to properly scan a building. Because of this, there were buildings which were not recognized. The fluctuating weather was also a challenge.

As mentioned before, the Product Reaction Cards method was applied after the user observation sessions, as part of the post-questionnaire method. The first 5 words which have dominated the usability evaluation for the AR app were: "Accessible", "Attractive", "Convincing", "Useful", "Intuitive".

The SUS score for the AR mobile app of Spotlight Heritage was 89.5, corresponding to the "Best Imaginable" usability adjective.

6. Discussion on the Overall Multi-Platform User Evaluation

6.1. General Discussion

Let us first look into some general positive and negative aspects of the usability of the digital components of Spotlight Heritage (SH). As we expected, the Interactive Touchscreen Table was appreciated for the enhanced view and the possibility of seeing the interface more clearly. The website evaluation session on the desktop/laptop and the evaluation of the SH mobile app and SH AR mobile app highlighted the ease of use and navigation but also the intuitiveness of the design. In addition, we found that the map is considered a very useful functionality for finding the landmarks of the project, both on the desktop website and on the SH mobile app. For the mobile apps, the users evaluated the matching colors to be on a positive note.

Very intriguing are the mixed feelings related to the possibility of adding your own memories about the project landmarks. Even if the idea is deemed very positive, increasing interactivity, credibility, and confidence, the implementation is considered to be the least intuitive section of the website.

Other common usability issues found on the website and the mobile app referred to the lack of a search bar, which is very important for users [28], and the similarity between the icons of some buttons.

For each of the chosen digital components for evaluation, there was an individual usability problem. For the website, it was the overcrowded menu, and for the Spotlight Heritage mobile app, it was the long loading time. For the Spotlight Heritage AR mobile app, the problems are based on the difficulties of positioning the camera in order to scan a building. Because of this, and also depending on weather (rain, strong sun, etc.), there were buildings which were not recognized, frustrating users. The authors are working on improving the building recognition algorithm for the SH AR mobile app [29,30].

Finally, the large dimensions of the Interactive Touchscreen Table bring some disadvantages, as not all buttons, menus, and keyboards are easily accessible from only one point, the user being required to move around the table.

Another discussion is related to the Product Reaction Cards, which we used for evaluating the website on desktop/laptop, but also for the two mobile applications. The most chosen words in the desktop/laptop evaluation were "Complex", "Accessible", "Understandable", "Efficient", and "Useful". The Spotlight Heritage mobile app was described as "Captivating", "Accessible", "Attractive", "Useful", and of "Good quality". Finally, the words most chosen in the Spotlight Heritage Augmented Reality mobile app evaluation were "Accessible", "Attractive", "Convincing", "Useful", and "Intuitive".

What we can conclude here is that all the digital products related to the Spotlight Heritage project are considered to be accessible and useful. The mobile applications are both rated as attractive. An interesting aspect is the fact that the website is most often described as Complex, even if the Spotlight Heritage mobile app has more or less the same information displayed. All words chosen in this evaluation method are positive ones, which proves to us that the digital components of this project were appreciated by the participants of the evaluation study.

Another intriguing aspect to compare is the SUS scores obtained by the different platforms (Table 5). The first remark is that all the four platforms received a very high SUS score, i.e., 89.5 to 92 out of 100, which is in the range of Excellent to Best Imaginable. This means that the digital components are highly usable, despite the usability problems that were described above.

Table 5. Comparison of average SUS scores for the four platforms (data collected during the usability testing sessions).

Platform	Average SUS Score	Adjective
Interactive touchscreen table	90	Best Imaginable
Desktop/Laptop	92	Best Imaginable
Mobile device	85.5	Excellent
Augmented Reality	89.5	Best Imaginable

The second remark is that the difference between them is very small, which suggests that users will not have big surprises when switching between platforms and will find the experience seamless and unitary.

However, the average SUS scores obtained through the online survey differ from the average SUS scores obtained during the usability testing sessions with as many as 10 points. In the online survey, the desktop/laptop version obtained 82.20 (compared to 92 in the usability testing sessions) and the mobile version obtained 79.19 (compared to 85 in the usability testing sessions).

An explanation could be that the usability testing sessions "gave" the participants some concrete tasks to perform in the application, which is consistent with scientific findings such as that "SUS scores are sensitive to successful task completion, with those completing tasks successfully providing higher scores" [31].

A complementary explanation is that participants in moderated usability testing sessions tend to please the moderator by giving higher scores [32] (p. 255).

Finally, a third explanation could be related to the fact that the online survey had many more older adults responding than similarly-aged people participating in the usability session, which is consistent with scientific literature which found out that younger adults tend to give better usability scores than older ones [33].

Regarding differences in SUS scores between men and women, we found a gap of 1.17 points for the desktop/laptop version (82.20 for men and 81.03 for women) and a gap of 4.76 for the mobile version (82.73 for men and 77.97 for women). While some studies suggest that gender plays a role in technology anxiety and adoption [34], others did not find significant differences in usability scores between men and women [35].

Concerning differences between respondents that had previously heard about or even used the Spotlight Heritage applications and respondents that were interacting with Spotlight Heritage for the first time, we did not find any significant difference in SUS scores (less than 0.5 points).

Lastly, both the desktop/laptop version and the mobile version got a very high score in the Net Promoter Score question in the online survey (68.09 for desktop/laptop and 70.45 for mobile), which means that the Spotlight Heritage scores very well in user satisfaction and loyalty.

Overall, the multi-platform evaluation indicates that the digital products are considered accessible and useful and that they score high in usability, across gender and age, which is an important factor in bringing people closer to cultural heritage.

6.2. Comparison with Similar Works

Compared to similar works described in Sections 2.1 and 4.1, our study addresses a more complex, multi-platform cultural concept, and uses a wider user evaluation methodology.

In ref. [5], the authors evaluate a museum mobile application with 43 participants using a combination of semi-structured interviews, focus group sessions and user observation sessions. In ref. [8], the authors evaluate an augmented reality app with olfactive extension, for an archaeological site, with 67 visitors, employing usability testing sessions, SUS questionnaires, and socio-demographic questionnaires. In ref. [9], the authors evaluate a cultural website using a quantitative assessment based on heuristic analysis metrics (it is not clear how many participants were involved in this part of their larger study). Finally, in [10], the researchers evaluate a 3D desktop application for a fortress with 6 experts from various fields.

In comparison with these works, our study evaluates a multi-platform system, composed of a interactive touchscreen table, desktop/laptop, mobile, and Augmented Reality platforms, with 105 participants, using a combination of semi-structured interviews, observations, think-aloud protocol, SUS questionnaire, Net Promoter Score, and Product Reaction Cards. This suggests a higher reliability of the results obtained.

Regarding the actual findings, there are similarities and differences between our study and these related works. For example, the researchers in [10] also emphasize the necessity of engaging professionals from multiple fields in the development of a storytelling cultural concept. Another example is that the researchers in [8] found no differences in the usability of the digital product when considering the sex of the participants, similarly to our results, but they also found no differences when considering the age, which was not the case in our study.

6.3. Threats to the Validity of the Study

Regarding the limitations of this evaluation, we acknowledge the existence of potential threats to the validity of our study. First of all, there is a selection threat to the validity of the moderated usability testing sessions, as the participants were not chosen randomly. Instead, a convenience-based sampling was used, i.e., our students probably recruited participants from their friends, colleagues, and acquaintances, although they were strongly advised not to do so. We attempted to alleviate this threat by also running the online survey evaluation, which gathered participants in a more random fashion.

Secondly, there is a risk that participants might have dropped interest in the experiment at some point, due to the high number of evaluation methods used in one session. We tried to alleviate this threat by arranging the methods in time in such a way that they feel natural and the session does not seem too fragmented. Additionally, the participants were instructed that they can leave the session at any point in time, without obligations or explanations.

Thirdly, another threat to the validity of our study is the population sample, which relates to the first acknowledged threat. We aimed to get participants with a representative demographic profile, but it proved to be difficult to engage professionals that work in other domains than ours. For example, the usability testing session participants mostly work in IT (10), while 6 are working in Education and 1 in Arts/Culture. As for the online survey respondents, 29 are teachers, 41 are students, and only 3 work in Public Administration. These numbers might affect the generalization of our study's results.

7. Conclusions

In this paper, we have described the technical design and implementation of a multi-platform digital storytelling concept called Spotlight Heritage, a 3 year project that we developed in Timisoara (Romania), European Capital of Culture in 2023. The digital architecture of the concept consists in applications which run on interactive touchscreen tables, desktop/laptop, mobile devices, and Augmented Reality.

Because of the lack of user evaluation models for such multi-platform digital storytelling concepts for cultural heritage, we have described here the user evaluation methodology, procedure, and results that we obtained after applying a mix of usability testing methods for each of the platforms and unified the insights.

For this purpose, we ran an online survey with 75 respondents in November 2020 and performed a set of in-person and remote moderated usability testing sessions with 30 participants in May and July 2021, using all-together a mix of methods consisting of semi-structured interviews, observations, think-aloud protocol, SUS questionnaire, Net Promoter Score question, and Product Reaction Cards.

We discovered that all the four platforms of the Spotlight Heritage concept have "Excellent" and "Best Imaginable" usability scores and that they are considered accessible and useful. This contributes to a seamless navigation between platforms and user satisfaction and loyalty towards the cultural heritage concept. We found that each platform complements the experience of the cultural heritage consumer: the interactive touchscreen table allows for social interaction around the cultural information, the desktop/laptop version provides the possibility of exploring deeply the content, the mobile platform allows for exploring the information on-the-go and sharing it with the others, and the Augmented Reality version provides the means for exploring the landmarks right in front of them.

We also identified some usability problems, such as the lack of a search button in the website, and in the mobile app, the effort needed to interact with certain areas of the application on the interactive touchscreen table or the difficulties in recognizing a landmark with the Augmented Reality application because of weather conditions.

The mix of evaluation methods that we used gave us a robust picture of the positive and negative aspects of the multi-platform usability and the results correlate, taken separately, with the scientific literature at hand.

We will use this research to improve the user experience of the multi-platform digital storytelling concept that we described here and as a basis for deeper research into user experience models for multi-platform concepts that include even newer technologies (mixed reality, outdoor holographic displays). Moreover, we intend to broaden the discussion of our findings in a future paper, on topics such as the comprehension and memorability of the digital storytelling techniques that we presented.

Author Contributions: Conceptualization, S.V. and D.A.; Data curation, S.V., O.R., M.M. and C.O.; Funding acquisition, R.V.; Methodology, S.V., D.A. and O.R.; Project administration, D.A.; Software, A.T. and C.O.; Supervision, R.V.; Validation, D.A. and V.M.; Visualization, A.T., O.R. and C.O.; Writing—original draft, S.V., V.M. and O.R.; Writing—review & editing, A.T. and M.M. All authors have read and agreed to the published version of the manuscript.

Funding: This research was funded by the Politehnica University of Timisoara under grant number 10162/11.06.2021.

Institutional Review Board Statement: The study was conducted according to the Ethical Regulations and Guidelines of the Politehnica University of Timisoara, Romania http://upt.ro/Informatii_etica-si-deontologie_164_ro.html (accessed on 30 July 2021).

Informed Consent Statement: GDPR data protection information was provided and respected and informed consent was obtained from all subjects involved in the study.

Data Availability Statement: The data presented in this study are available on request from the corresponding author.

Acknowledgments: We gratefully acknowledge the contributions of all participants in our experiments, as well as those involved in the Spotlight Heritage Timisoara cultural project.

Conflicts of Interest: The authors declare no conflict of interest.

References

1. Tiriteu, D.; Vert, S. Usability Testing of Mobile Augmented Applications for Cultural Heritage—A Systematic Literature Review. In Proceedings of the RoCHI 2020, Sibiu, Romania, 22–23 October 2020; p. 144. [CrossRef]
2. De Jager, A.; Fogarty, A.; Tewson, A.; Lenette, C.; Boydell, K.M. Digital storytelling in research: A systematic review. *Qual. Rep.* **2017**, *22*, 2548–2582.
3. Wu, J.; Chen, D.-T.V. A systematic review of educational digital storytelling. *Comput. Educ.* **2020**, *147*, 103786. [CrossRef]

4. Lambert, J. *Digital Storytelling: Capturing Lives, Creating Community*, 4th ed.; Routledge: New York, NY, USA; London, UK, 2012.
5. Katifori, A.; Tsitou, F.; Pichou, M.; Kourtis, V.; Papoulias, E.; Ioannidis, Y.; Roussou, M. Exploring the Potential of Visually-Rich Animated Digital Storytelling: The Mobile Experience of the Athens University History Museum. In *Visual Computing for Cultural Heritage*; Springer International Publishing: Berlin/Heidelberg, Germany, 2020; p. 325.
6. Ivey, B. *Arts, Inc.: How Greed and Neglect Have Destroyed Our Cultural Rights*; University of California Press: Berkeley, CA, USA, 2010.
7. Ioannidis, Y.; El Raheb, K.; Toli, E.; Katifori, A.; Boile, M.; Mazura, M. One object many stories: Introducing ICT in museums and collections through digital storytelling. In Proceedings of the 2013 Digital Heritage International Congress (DigitalHeritage), Marseille, France, 28 October–1 November 2013; Volume 1, pp. 421–424.
8. Marto, A.; Melo, M.; Gonçalves, A.; Bessa, M. Development and Evaluation of an Outdoor Multisensory AR System for Cultural Heritage. *IEEE Access* **2021**, *9*, 16419–16434. [CrossRef]
9. Psomadaki, I.; Dimoulas, C.A.; Kalliris, G.M.; Paschalidis, G. Technologies of Non Linear Storytelling for the Management of Cultural Heritage in the Digital City: The Case of Thessaloniki. In *Digital Cultural Heritage*; Springer: Berlin/Heidelberg, Germany, 2018; pp. 337–349.
10. Rizvic, S.; Djapo, N.; Alispahic, F.; Hadzihalilovic, B.; Cengic, F.F.; Imamovic, A.; Okanovic, V.; Boskovic, D. Guidelines for interactive digital storytelling presentations of cultural heritage. In Proceedings of the 2017 9th International Conference on Virtual Worlds and Games for Serious Applications (VS-Games), Athens, Greece, 6–8 September 2017; pp. 253–259.
11. Fernandez, A.; Insfran, E.; Abrahão, S. Usability evaluation methods for the web: A systematic mapping study. *Inf. Softw. Technol.* **2011**, *53*, 789–817. [CrossRef]
12. Koukopoulos, Z.; Koukopoulos, D. Evaluating the Usability and the Personal and Social Acceptance of a Participatory Digital Platform for Cultural Heritage. *Heritage* **2019**, *2*, 1. [CrossRef]
13. Ahmad, N.A.N.; Lokman, A.; Ab Hamid, N.I.M. Performing Usability Evaluation on Multi-Platform Based Application for Efficiency, Effectiveness and Satisfaction Enhancement. *Int. J. Interact. Mob. Technol.* **2021**, *15*, 103–117. [CrossRef]
14. Grier, R.A.; Bangor, A.; Kortum, P.; Peres, S.C. The System Usability Scale: Beyond Standard Usability Testing. In Proceedings of the Human Factors and Ergonomics Society Annual Meeting, Torino, Italy, 1 September 2013; Volume 57, pp. 187–191. [CrossRef]
15. Sauro, J.; Lewis, J. When designing usability questionnaires, does it hurt to be positive? In Proceedings of the SIGCHI Conference on Human Factors in Computing Systems, Vancouver, BC, Canada, 7–12 May 2011; p. 2224. [CrossRef]
16. Sauro, J.; Lewis, J.R. Chapter 8—Standardized usability questionnaires. In *Quantifying the User Experience*, 2nd ed.; Sauro, J., Lewis, J.R., Eds.; Morgan Kaufmann: Boston, MA, USA, 2016; pp. 185–248. [CrossRef]
17. Das, A.; Acar, G.; Borisov, N.; Pradeep, A. The Web's Sixth Sense: A Study of Scripts Accessing Smartphone Sensors. In Proceedings of the 2018 ACM SIGSAC Conference on Computer and Communications Security, New York, NY, USA, 15–19 October 2018; pp. 1515–1532. [CrossRef]
18. Adams, W. Conducting Semi-Structured Interviews. In *Handbook of Practical Program Evaluation*; John Wiley & Sons, Ltd.: Hoboken, NJ, USA, 2015; pp. 492–505. [CrossRef]
19. Bastien, J.M.C. Usability testing: A review of some methodological and technical aspects of the method. *Int. J. Med. Inf.* **2010**, *79*, e18–e23. [CrossRef] [PubMed]
20. Krahmer, E.; Ummelen, N. Thinking About Thinking Aloud: A Comparison of TwoVerbal Protocols for Usability Testing. *IEEE Trans. Prof. Commun.* **2004**, *47*, 105–117. [CrossRef]
21. Mclellan, S.; Muddimer, A.; Peres, S. The Effect of Experience on System Usability Scale Ratings. *J. Usability Stud.* **2012**, *7*, 56–67.
22. Bangor, A.; Kortum, P.; Miller, J. Determining what individual SUS scores mean: Adding an adjective rating scale. *J. Usability Stud.* **2009**, *4*, 114–123.
23. Keiningham, T.; Aksoy, L.; Cooil, B.; Andreassen, T.; Williams, L. A Holistic Examination of Net Promoter. *J. Database Mark. Cust. Strategy Manag.* **2008**, *15*, 79–90. [CrossRef]
24. Barnum, C.M. 7—Conducting a usability test. In *Usability Testing Essentials*, 2nd ed.; Barnum, C.M., Ed.; Morgan Kaufmann: Boston, MA, USA, 2021; pp. 249–285. [CrossRef]
25. Andone, D.; Vert, S.; Mihaescu, V.; Stoica, D.; Ternauciuc, A. Evaluation of the Virtual Mobility Learning Hub. In *Learning and Collaboration Technologies. Designing, Developing and Deploying Learning Experiences*; Springer Nature: Cham, Switzerlands, 2020; pp. 20–33. [CrossRef]
26. Vert, S.; Andone, D. Mobile usability evaluation: The case of the Art Encounters 2017 application. In Proceedings of the International Conference on Human-Computer Interaction (RoCHI 2018), Cluj-Napoca, Romania, 3–4 September 2018; pp. 42–45.
27. Jung, M.; Lazaro, M.J.S.; Yun, M.H. Evaluation of Methodologies and Measures on the Usability of Social Robots: A Systematic Review. *Appl. Sci.* **2021**, *11*, 1388. [CrossRef]
28. Hur, Y.; Jo, J. Development of Intelligent Information System for Digital Cultural Contents. *Mathematics* **2021**, *9*, 238. [CrossRef]
29. Orhei, C.; Vert, S.; Vasiu, R. A Novel Edge Detection Operator for Identifying Buildings in Augmented Reality Applications. In *Information and Software Technologies*; Elsevier: Cham, Switzerlands, 2020; pp. 208–219. [CrossRef]
30. Orhei, C.; Vert, S.; Mocofan, M.; Vasiu, R. End-To-End Computer Vision Framework: An Open-Source Platform for Research and Education. *Sensors* **2021**, *21*, 3691. [CrossRef] [PubMed]
31. Lewis, J.R.; Sauro, J. Item benchmarks for the system usability scale. *J. Usability Stud.* **2018**, *13*, 158–167.
32. Barnum, C. *Usability Testing Essentials: Ready, Set...Test!* 2nd ed.; Morgan Kaufmann: Boston, MA, USA, 2020.

33. Sonderegger, A.; Schmutz, S.; Sauer, J. The influence of age in usability testing. *Appl. Ergon.* **2016**, *52*, 291–300. [CrossRef]
34. Schmidt, M.; Kafka, J.X.; Kothgassner, O.D.; Hlavacs, H.; Beutl, L.; Felnhofer, A. Why Does It Always Rain on Me? Influence of Gender and Environmental Factors on Usability, Technology Related Anxiety and Immersion in Virtual Environments. In Proceedings of the 10th International Conference on Advances in Computer Entertainment—Volume 8253, Boekelom, The Netherlands, 12–15 November 2013; pp. 392–402.
35. Bangor, A.; Kortum, P.T.; Miller, J.T. An Empirical Evaluation of the System Usability Scale. *Int. J. Human Comput. Interact.* **2008**, *24*, 574–594. [CrossRef]

Article

RFaNet: Receptive Field-Aware Network with Finger Attention for Fingerspelling Recognition Using a Depth Sensor

Shih-Hung Yang [1,*,†], Yao-Mao Cheng [2,†], Jyun-We Huang [1] and Yon-Ping Chen [2]

1. Department of Mechanical Engineering, National Cheng Kung University, Tainan City 701, Taiwan; z10806026@ncku.edu.tw
2. Institute of Electrical Control Engineering, National Yang Ming Chiao Tung University, Hsinchu City 300, Taiwan; mark228926.ee08@nycu.edu.tw (Y.-M.C.); ypchen@cc.nctu.edu.tw (Y.-P.C.)
* Correspondence: vssyang@gs.ncku.edu.tw; Tel.: +886-6-27-57575 (ext. 62171)
† These authors contributed equally to this paper.

Abstract: Automatic fingerspelling recognition tackles the communication barrier between deaf and hearing individuals. However, the accuracy of fingerspelling recognition is reduced by high intra-class variability and low inter-class variability. In the existing methods, regular convolutional kernels, which have limited receptive fields (RFs) and often cannot detect subtle discriminative details, are applied to learn features. In this study, we propose a receptive field-aware network with finger attention (RFaNet) that highlights the finger regions and builds inter-finger relations. To highlight the discriminative details of these fingers, RFaNet reweights the low-level features of the hand depth image with those of the non-forearm image and improves finger localization, even when the wrist is occluded. RFaNet captures neighboring and inter-region dependencies between fingers in high-level features. An atrous convolution procedure enlarges the RFs at multiple scales and a non-local operation computes the interactions between multi-scale feature maps, thereby facilitating the building of inter-finger relations. Thus, the representation of a sign is invariant to viewpoint changes, which are primarily responsible for intra-class variability. On an American Sign Language fingerspelling dataset, RFaNet achieved 1.77% higher classification accuracy than state-of-the-art methods. RFaNet achieved effective transfer learning when the number of labeled depth images was insufficient. The fingerspelling representation of a depth image can be effectively transferred from large- to small-scale datasets via highlighting the finger regions and building inter-finger relations, thereby reducing the requirement for expensive fingerspelling annotations.

Keywords: fingerspelling recognition; depth sensor; finger attention; receptive field; inter-finger relation

1. Introduction

For deaf people, sign language is a means to communicate. However, communication between deaf and hearing people remains challenging. Automatic sign language recognition tackles this communication barrier by translating sign language to text or speech. Fingerspelling is a sign language that signals words letter by letter. Fingerspelling enables the communication of technical terms and other terms lacking a representation in sign language. Note that ~35% of words in social interactions refer to technical topics requiring fingerspelling [1].

Vision-based fingerspelling recognition has been widely developed because cameras are inexpensive and ubiquitously available. Fingerspelling recognition systems may benefit from depth images acquired by structured light or time-of-flight sensors, which are robust to illumination variations [2] and enable easy hand detections against a complex background. However, intra-class variability, inter-class similarity, and inter-subject variability hinder vision-based fingerspelling recognition, as shown in Figure 1. The inter-class similarity refers to different fingerspelling signs sharing similar hand postures. The intra-class variability refers to the various representations of identical signs captured from multiple views. The inter-subject variability refers to the various representations of identical signs performed by different subjects.

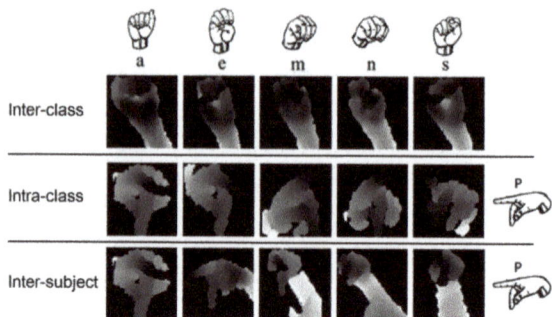

Figure 1. Inter-class similarity, intra-class variability, and inter-subject variability of hand gestures captured by a depth camera [3].

Accurate finger localization can potentially tackle inter-class similarity by detecting small hand posture variations between two signs. To account for finger localization, we propose a finger attention mechanism that enhances the finger regions in the depth image, highlighting discriminative finger features in these regions. Moreover, building inter-finger relations can tackle intra-class variability because inter-finger relations are inherently invariant to viewpoint changes. To account for inter-finger relations, we enlarge the receptive fields (RFs) of the convolutional kernels and model neighboring and inter-region dependencies between fingers. We now present the challenges of inter-class similarity and intra-class variability and illustrate our approach to handling them.

The first challenge is caused by inter-class similarity. A standard way to handle this is to highlight finger regions because accurate finger localization facilitates identifying slight posture variations across signs and further distinguishing signs that show inter-class similarity. The conventional method [4] assumes that finger localization and finger occlusion problems can be solved by the depth level. For this purpose, it manually decomposes the hand image into a few depth levels. Finger localization by this method is affected not only by manual predefinition of the depth level but also by the appearance of the forearm region at the depth level of the fingers. To deal with this issue, we highlighted the finger regions for low-level feature extraction using *depth finger attention* (DFA), as shown in Figure 2. DFA simultaneously considers the hand depth image and the forearm-removed image (non-forearm image) to explore the finger regions. There are two reasons behind simultaneously considering the hand depth image and non-forearm image. The first reason is that as most subjects signal with a preferred posture, the background pattern is sign-dependent. A model may tend to learn the sign representation according to the background pattern, which biases the learning toward the background [5]. When the fingers inside the hand region are not highlighted, the finger localization can be incorrect, and signs with inter-class similarity are poorly recognized. To handle the background-bias problem, we provide the non-forearm image as a reference for the hand depth image. The model then highlights the inside of the hand region rather than the outside region (the sign-dependent background pattern). The second reason is interference by forearm appearance in the hand depth image. Although the forearm can be removed from the hand image by detecting the wrist point [6], this approach is non-robust to occlusion of the wrist point by the fingers. Such occlusions can lead to inaccurate wrist point detection and unexpected finger removal. To tackle this limitation, we provide the hand depth image containing the forearm as a complementary reference for the non-forearm image. DFA can then highlight the finger regions on at least one of the two images.

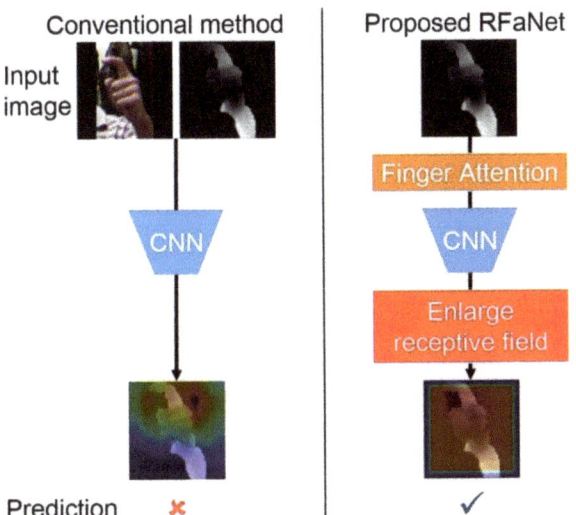

Figure 2. Hand gesture recognition models: (**left**) the conventional method and (**right**) RFaNet. Most conventional methods recognize hand gestures in color and depth input images, whereas RFaNet only processes a depth image to recognize hand gestures. RFaNet employs finger attention to highlight fingers before a CNN extracts the features and enlarges the RF to build long-range connections across finger features for better hand-gesture recognition.

The second challenge is caused by intra-class variability over multiple views. An identical sign viewed from multiple angles can have various representations in a convolutional neural network (CNN), leading to poor recognition. As convolutional operations focus on the local neighborhood, they capture the local finger features within a small RF (or field of view [7]), as shown in in Figure 2. When one sign is viewed from a different perspective, the change in local finger features leads to a variant representation. However, the long-range dependency between the fingers of an identical sign is invariant to viewpoint changes. Capturing the long-range dependency could improve the recognition of signs with intra-class variability. To handle this issue, we designed the second key component, a *non-local receptive field* (NLRF), that captures the neighboring and inter-region dependencies between fingers. The NLRF block employs atrous spatial pyramid pooling (ASPP) [7] to enlarge the field of view on multiple scales, and hence develops the long-range dependencies of distant fingers, as shown in Figure 2. Although ASPP varies the sampling distance from the kernel center, the feature maps from the previous convolutional layer have a uniform resolution. Consequently, the background enhancement is incorrect and the features are rendered less discriminative. Inspired by the receptive field block [8], we employed standard convolutional operations with various kernel sizes followed by the atrous convolution, accounting for the impact of RF eccentricities. However, directly merging the feature maps from various kernel sizes into a spatial pooling may model the dependency between fingers and the neighboring background rather than the dependency between distal fingers. The neighboring and inter-region dependencies are not simultaneously considered. To avoid this problem, we modified the non-local block [9] to further capture the dependencies of the feature maps extracted from various RFs. The non-local operation computed interactions between the multi-scale feature maps, and thus jointly captured the neighboring and inter-region dependencies across distal fingers, facilitating the modeling of inter-finger relations. Because the inter-finger relations of a sign are inherently invariant to viewpoint changes, a representation based on inter-finger relations could reduce intra-class variability.

Fingerspelling recognition systems may experience limited accuracy when the number of labeled images is insufficient. The number of labeled data can be increased by inviting multiple subjects to perform hand gestures under various conditions, but this approach is expensive. Furthermore, the data annotation of hand gestures often requires specialized domain knowledge, which reduces the scalability of the data. Transfer learning tackles this issue by training a deep neural network model via sufficiently many data in a source domain and fine-tuning the model using small data in a target domain [10]. The source domain does not necessarily require relevance to the target domain but must share certain common representations with it. The representations learned from the large-scale datasets facilitate learning from the small-scale datasets. Nihal *et al.* [11] observed that computer-vision tasks share similar features. They trained a model on ImageNet and transferred the knowledge to Bangla sign alphabet recognition [12]. Observing similar hand gestures in British and American sign languages, Bird *et al.* [13] conducted transfer learning from British to American sign languages, based on color modality and bone modality (finger joints). However, the background of the color modality may affect transfer learning in this method. The depth modality could facilitate the transfer learning of fingerspelling recognition because finger features are robust to illumination changes and background complexity. Therefore, in this study, we only adopted depth modality for fingerspelling recognition and demonstrate its advantage in the application of transfer learning on limited training datasets.

The DFA and NLRF blocks were the key components for mitigating inter-class similarity and intra-class variability, respectively. We assembled the DFA and NLRF blocks to the top and bottom of a backbone network (VGG-9 [14]) and proposed a model—*Receptive Field-aware Network with finger attention* (RFaNet)—for fingerspelling recognition, as shown in Figure 3. The primary contributions of the proposed model to fingerspelling recognition are summarized below.

1. We introduce RFaNet for effective fingerspelling recognition.
2. The DFA block on the top of RFaNet highlights the finger regions and facilitates the identification of slight hand-posture variations across signs with inter-class similarity.
3. The NLRF block at the bottom of RFaNet captures inter-finger relations by fusing multi-scale feature maps of various RFs. By learning the representations of inter-finger relations, the NLRF block improves the recognition of signs with intra-class variability because the representation of a sign is invariant to viewpoint changes.
4. RFaNet outperformed state-of-the-art methods on two standard benchmark fingerspelling datasets.
5. RFaNet effectively learned the fingerspelling representations from large- to small-scale datasets by highlighting the finger regions when the training data were insufficient.

The rest of the paper is organized as follows: Section 2 presents a review of related works in the literature; Section 3 describes RFaNet for fingerspelling recognition; Section 4 presents the experimental results, which are compared and analyzed; Section 5 extensively describes the experimental results of the RFaNet in transfer learning applications; and Section 6 concludes the study.

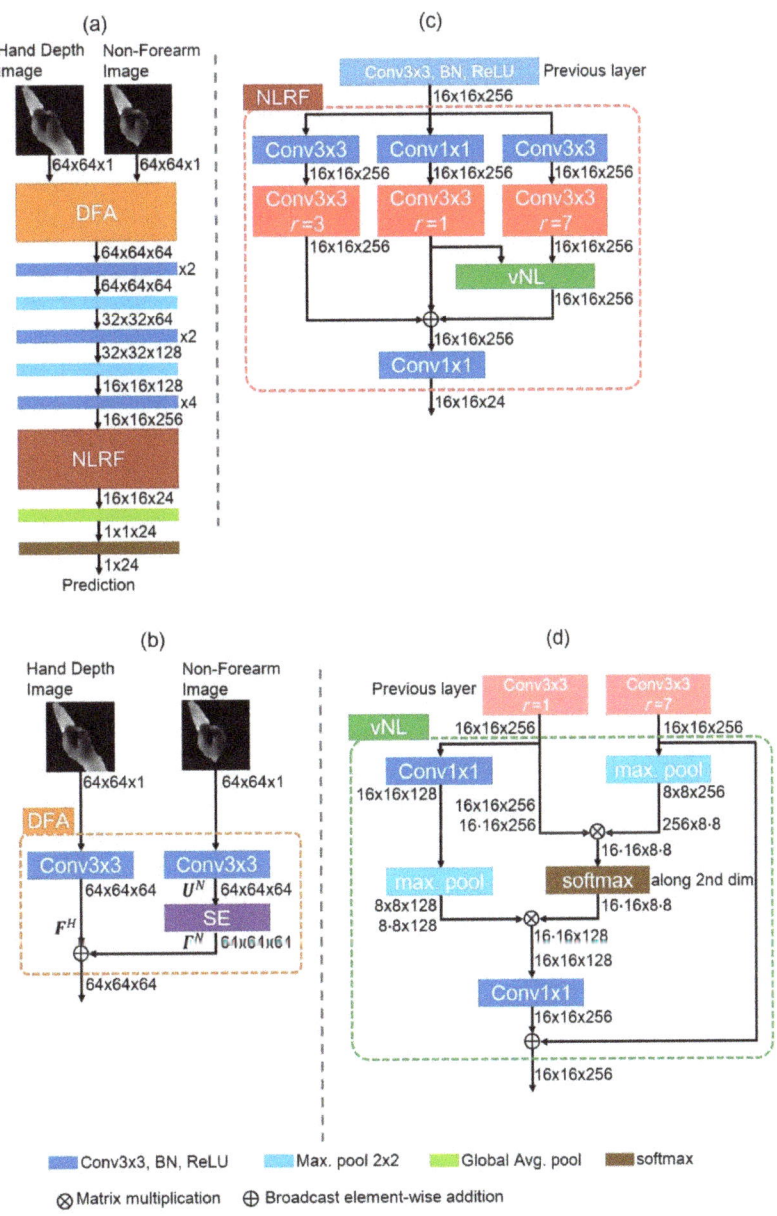

Figure 3. Overview of our hand gesture recognition model. (**a**) RFaNet. (**b**) DFA block. (**c**) NLRF block. (**d**) Variant of non-local block (vNL). SE: squeeze-and-excitation block; non-forearm image: hand depth image after forearm removal process; r: atrous sampling rate which corresponds to the stride when sampling the input signal.

2. Related Work

In this section, we describe the relevant recent works on fingerspelling recognition, RF, and attention mechanisms.

2.1. Fingerspelling Recognition

Usually, fingerspelling recognition applies depth modality, which is robust to illumination variations. Hu et al. [15] detected hands by assuming them as the closest objects to the sensor in depth images. Zhang and Tian [16] extracted the depth features and integrated them with a three-dimensional point cloud. Wang et al. [17] considered not only the depth-modality features (depth and skeleton features) but also the color modality features (color, texture, and contour features). Tao et al. [18] applied a CNN that recognizes letter signs captured from different perspectives in the depth modality. Modanwal and Sarawadekar [6] observed that the forearm usually appears in the hand image and is irrelevant to the hand gesture. They suggested removing the forearm from the hand image to improve hand gesture recognition. They developed a robust wrist-point detection algorithm to separate the palm and forearm based on hand anatomy. Removing the forearm is essential for capturing the fingers and extracting finger features in the hand image. Motivated by this result, we removed the forearm from the hand-depth image and extracted the low-level features from the finger and palm regions.

Rioux-Maldague and Giguère [4] decomposed the depth map of the hand into several layers, each representing a depth-level of the hand region. Partial fingers and palm regions at similar depth values appear in the same layer and are represented as depth features at the corresponding depth level. Decomposing a hand into different depth levels can handle finger occlusion. When one finger partially occludes another finger, both fingers belong to two depth levels and appear in two layers. This facilitates the localization of fingers, which is important for distinguishing fingerspelling signs. Accordingly, we were motivated to decompose the hand region into several depth-feature maps containing various depth information and facilitating finger localization.

2.2. Receptive Field

Conventional methods usually employ very deep convolutional networks that recognize objects at multiple scales, leading to huge computational costs. The cost can be reduced by replacing deep backbones with a lightweight model, in which enlarged RFs can potentially increase the field of view at multiple scales. The ASPP [7] enlarges the RFs by changing the sampling distance from the kernel center to capture the long-range dependency. Using ASPP, Wang et al. [19] extracted the spatial information around an object occluded by other objects. ASPP exploits and preserves the fine details around occlusions. Tan et al. [20] yielded a fixed-length feature representation using spatial pyramid pooling, which recognizes hand gestures regardless of input size. This method facilitates the propagation of gradients from the final fully connected layer to the input layer. The resolution of the input feature maps from the previous convolutional layers is uniform in the ASPP. Lu et al. [21] suggested that when inferring occlusion relationships, a sufficient RF is required at different scales for aggregating the cues around the occlusion region. Therefore, they extended the ASPP to different scales of the RF, enabling the complete sensing of foreground and background objects. Liu et al. [8] developed a receptive field block (RFB) that considers the relationship between the size and eccentricity of the RF. The RFB improves feature representation and can be equipped on top of a lightweight network for object detection tasks.

2.3. Attention Mechanism

Attention mechanisms are helpful for recalibrating the channel dependency of a computer vision task [22]. They model the long-range dependency of natural language processing [23]. Wang et al. [24] designed a residual ASPP block that extracts multiscale features from stereo images and a parallax-attention module that fuses these multiscale features to capture the stereo correspondences. Han et al. [25] simultaneously applied an ASPP block and a channel attention module for multiscale context extraction and channel-wise feature recalibration, respectively. The features extracted from the two branches were fused by weighted summation for the semantic labeling of high-resolution remote sensing images. Liu et al. [26] densely connected the branches of an ASPP to cover the dense feature scales of RGB and depth modalities. Using a selective self-mutual attention module, they then integrated the attentions of the RGB and depth modalities to capture the long-range dependencies in RGB-D salient object detection. Yang et al. [27] developed a depth-aware attention module to refine the RGB and depth feature maps for suppressing the effect of color–depth misalignment. This module highlights important fingers for fingerspelling recognition. Inspired by the interactive learning of attentions from two modalities, we exploited the merits of ASPP and attention mechanisms to enlarge RFs at multiple scales and build the long-range dependencies of distant fingers. Our idea is to leverage the neighboring and inter-region neighboring dependencies between fingers. The resulting fingerspelling representation is invariant to viewpoint changes and further reduces intra-class variability.

3. Receptive Field-Aware Network with Finger Attention

In this section, we first introduce the overall architecture of the proposed fingerspelling recognition method, RFaNet; then, we describe how the key components of RFaNet facilitate tackling the fingerspelling recognition tasks.

Figure 3a shows the overall architecture of RFaNet. RFaNet was trained to enhance the finger regions and to build inter-finger relations in the depth image. A VGG-9 [14] is adopted as the backbone network. The proposed DFA and NLRF blocks are inserted at the top and bottom of RFaNet, respectively. The DFA block was designed to extract the low-level features from the finger and palm regions rather than the background regions. The NLRF block is designed to fuse the neighboring and inter-region information and extract a fingerspelling representation invariant to viewpoint changes. Experimental results supported the hypothesis that the DFA and NLRF blocks improved the overall fingerspelling recognition performance. We share our code and models at: https://github.com/yaomao-cheng/RFaNet_model/tree/master (accessed on 25 October 2021).

3.1. Depth Finger Attention Block

Fingerspelling recognition is usually hindered by inter-class similarity, i.e., by the similar appearances of more than one sign. Accurate finger localization is crucial for identifying slight hand posture variations. Unlike the method in [4], which manually divides the hand depth image into several depth-level layers for finger localization, the proposed DFA block applies learnable convolutional operations to obtain several depth feature maps from a hand depth image. However, the convolutional model may tend to learn sign-dependent background patterns, because most subjects make signs with a preferred posture, resulting in similar background patterns for identical signs (known as the background-bias phenomenon [5]). To guide the model toward the finger regions, the DFA block jointly processes two depth images: the hand depth image and the same image with the forearm removed (non-forearm image), which provide complementary information, as shown Figure 3b. As the non-forearm image references the hand depth image, the DFA block can highlight inside the hand region rather than the outside, i.e., a sign-dependent background pattern. The forearm was removed by a wrist-point detection algorithm [6], thus creating the non-forearm images. However, when the fingers occlude the wrist point, they can be incorrectly removed by the algorithm. In such cases, the hand depth image (possessing forearm) could provide complementary finger information that enhances the finger region.

In the non-forearm image path, the squeeze-and-excitation (SE) block [22] is employed to adaptively recalibrate relations across feature maps to effectively highlight finger regions, as shown in Figure 3b. These recalibrated feature maps from the non-forearm image are fused (by addition) with the feature maps from the hand depth image for learning to excite finger regions. This fusion ensures that the hand depth image and non-forearm image could provide complementary finger features, leading to finger localization even under wrist occlusion. The DFA block is inserted in the first layer of the proposed model, which enables the following layers to extract discriminative features in the finger regions, as shown in Figure 3a. It facilitates the identification of slight hand posture variation across signs with inter-class similarity.

Given a feature map of the non-forearm image $U^N = [u_1^N, u_2^N, \ldots, u_C^N] \in \mathbb{R}^{H \times W \times C}$ extracted by the convolutional kernels, where $H = W = C = 64$ in this study, the SE block first squeezes the global spatial information via a global average pooling to obtain the channel-wise statistics, as follows:

$$z_c^N = \frac{1}{H \times W} \sum_{i=1}^{H} \sum_{j=1}^{W} u_c^N(i,j), \tag{1}$$

where z_c^N represents the channel-wise statistics of the c-th channel. Then, the SE block captures channel-wise dependencies by two fully connected layers, as follows:

$$s^N = \sigma\left(W_2 \delta\left(W_1 z^N\right)\right), \tag{2}$$

where σ and δ denote the sigmoid activation and rectified linear unit [28] functions, respectively, $W_1 \in \mathbb{R}^{\frac{C}{r} \times C}$, and $W_2 \in \mathbb{R}^{C \times \frac{C}{r}}$. We set the reduction ratio r to 2. The output of the SE block was obtained by recalibrating the channel-wise features, as follows:

$$F^N = s^N \otimes U^N, \tag{3}$$

where $\otimes$ represents the element-wise product implemented by broadcasting the s^N values along the spatial axis. The SE block learns to excite the informative features of the non-forearm image and can potentially boost the finger localization ability.

The DFA block fuses the feature maps of the hand depth image F^H and the non-forearm image F^N. Among several operations in the fusion strategy—addition, product, and concatenation—we empirically reported that the addition operation provides better classification accuracy at less computational cost than the others. Therefore, addition was selected as the fusion strategy of F^H and F^N in the DFA block. By recalibrating the channel-wise dependencies of the features F^N, the DFA exploits the contextual information outside small RFs and enhances the features inside the hand region. The feature map F^N provides a reference for F^H, guiding the model toward the finger regions rather than sign-dependent background patterns. Moreover, the feature map F^H provided complementary information to F^N when the fingers were incorrectly removed in F^N under wrist occlusion. Jointly processing F^H and F^N focuses the attention on fingers in the depth image by highlighting the salient finger regions, thus improving the low-level finger representations.

3.2. Non-Local Receptive Field Block

A fingerspelling sign captured from multiple views may have various representations, resulting in intra-class variability. However, the inter-finger relations of a sign are inherently invariant to viewpoint changes. To capture inter-finger relations, the proposed NLRF block enlarges the RF and field of view to capture the long-range dependencies of distal fingers. Unlike the ASPP [7] and receptive field block (RFB) [8], the NLRF block not only applies standard convolutional operations with various receptive fields, followed by the atrous convolution, but also modifies the non-local block [9] to capture the relation of feature maps with multiple fields of view, which facilitates the modeling of the relations between distal fingers. The NLRF block exploits multi-scale feature maps using three atrous convolutions, with rates $r = 1, 3$, and 7, as shown in Figure 3c. The reason behind using the three rates is that the atrous convolution with a high rate only samples a region with checkerboard patterns, leading to a gridding problem [29] and the loss of neighboring information. We followed the suggestion in [29] to select rates that did not possess a common factor relationship (i.e., 1, 3, and 7). The rate parameter r represents the stride where the operator sampled the input signal. We applied the maximal atrous sampling rate $r = 7$ because the feature map from the previous layer is of spatial resolution 16 × 16. We empirically found that the atrous convolution with $r = 5$ did not significantly improve the classification accuracy, and thus was removed (see Section 4.5 for a detailed analysis). The removal of the atrous convolution with $r = 5$ reduces the computational cost.

To relate the small and large fields of view, the feature maps from the branches of atrous convolution with rates $r = 1$ and 7 are fused by a variant of non-local (vNL) block. The reason underlying this fusion step is shown in Figure 4. The ASPP and RFB directly merge the feature maps with various RFs and fields of view, such that all pixels in the spatial array of RF equally contribute to the output response. Therefore, the relation between finger and background may be modeled rather than that between distal fingers (e.g., index finger and thumb), leading to incomplete inter-finger relations. The branch $r = 1$ captures neighboring information in a local area, whereas the branch $r = 7$ captures inter-region information in a large area. The neighboring information could provide the local relations between neighboring fingers, whereas the inter-region information could provide the non-local relations between distal fingers. Fusing the neighboring and inter-region information could emphasize the most essential regions, according to local and non-local relations, and better model inter-finger relations, as shown in Figure 4.

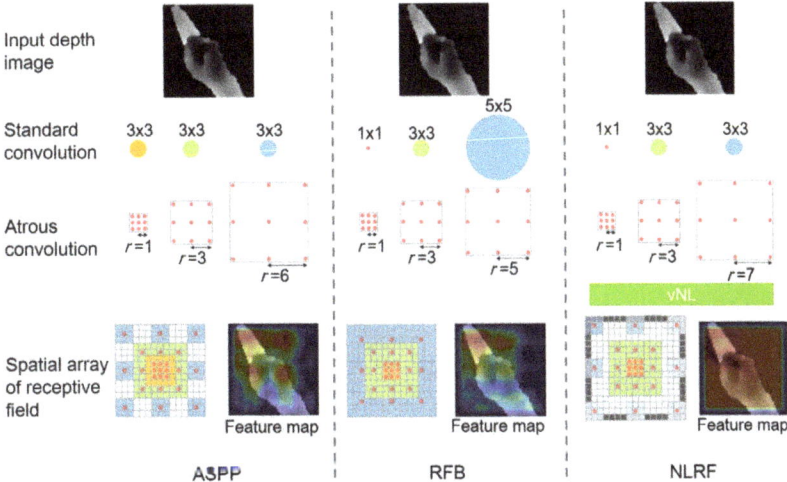

Figure 4. We adjusted the RF sizes in the original ASPP and RFB for a fair comparison. The feature maps are derived from the output of ASPP, RFB, and NLRF. The dark outer region of the feature map of the NLRF is zero-padded to fit the RF of atrous convolution with rate $r = 7$. vNL: variant of non-local block; r: atrous sampling rate.

The non-local block [9] applies a self-attention mechanism to enhance the features at a given position by aggregating the information at other positions of the same input feature vector. Different from the non-local block, which derives the value, key, and query from an identical input, our vNL enhances the features at a position of the atrous convolution with the rate $r = 7$ (a large RF) by aggregating the information at other positions of atrous convolution with the rate $r = 1$ (a small RF). The vNL facilitates the modeling of the long-range dependencies of multi-scale feature maps. Figure 3d shows that the vNL processes the feature maps produced from the branches $r = 1$ and 7 in the previous layer. The vNL shares a similar framework to the non-local block comprising context modeling, transformation, and fusion [30]. The global context features are modeled as the dot-product (matrix multiplication) of the feature embeddings of two positions, in the branches $r = 1$ and 7, respectively. The channel-wise dependencies of the global context features, captured by a 1×1 convolution, are as shown at the bottom of Figure 3d. The global context features are aggregated at the features of each position in the branch $r = 7$ by a broadcast element-wise addition. We employed max-pooling to the feature maps of the branches $r = 1$ and 7 after a linear transformation to reduce the computational cost and extract shift-invariant features. The max-pooling could reduce the background effect because the background feature values were smaller than the hand feature values. Figure 4 shows that the NLRF block effectively captures the neighboring dependency of the index and middle fingers in the small RF and the inter-region dependency of the index finger and thumb within the large RF.

3.3. Optimization

Optimization was performed by summing two loss functions. The first loss function is the categorical cross-entropy loss function for multi-class classification:

$$\mathcal{L}_{CE} = -\frac{1}{NK} \sum_{i=1}^{N} \sum_{k=1}^{K} y_{ik} \log \hat{y}_{ik}, \quad (4)$$

where N is the mini-batch size, K is the number of classes, y_{ik} denotes the ground-truth label, and $\hat{y}_{ik}$ is the network output.

We also considered the sparsity-induced penalty term [31] in the loss function. This penalty term forces the scaling factors to be sparse in the batch normalization layer to improve the generalization ability. The complete loss for training RFaNet is as follows:

$$\mathcal{L} = \mathcal{L}_{CE} + \lambda \sum_{\gamma \in \Gamma} |\gamma|, \quad (5)$$

where γ is the scaling factor, Γ is the set of scaling factors in the network, and λ regulates the tradeoff between the classification accuracy and generalization ability.

4. Experimental Results

4.1. Datasets

We evaluated RFaNet on the following datasets. Each sample in these datasets consists of a pair of RGB and depth images. Figure 5 shows sample depth images from these datasets, where certain signs share similar hand shapes.

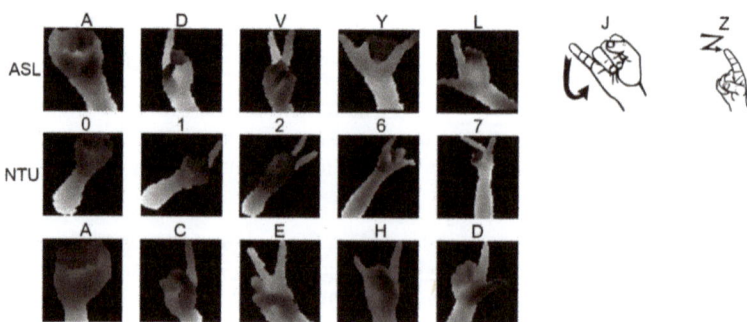

Figure 5. Samples from ASL, NTU, and OUHANDS datasets. The letter on the top of each panel represents the gesture label of the dataset. Each column represents the gestures that share similar hand shapes but different labels across three datasets. The letters *j* and *z* were excluded due to their dynamic characteristics.

ASL Fingerspelling Dataset. The ASL fingerspelling dataset comprises 24 letter signs of the American Sign Language alphabet acquired by the Microsoft Kinect sensor [3]. The dynamic letters *j* and *z* were excluded because RFaNet recognizes fingerspelling from a single depth image, which cannot reveal the dynamic characteristics of the letters, as shown in Figure 5. These letter signs were performed by five subjects in front of various backgrounds and from different viewpoints. Each letter sign has 500 samples for each subject.

This dataset contains a few invalid samples in which the hand is missing or the letter sign does not belong to the ground-truth label, which was described in Yang et al. [27]. Therefore, we manually removed these invalid samples from the training and testing data.

NTU Digit Dataset. The NTU digit dataset comprises 10 digit signs acquired by the Microsoft Kinect sensor [32]. These digit signs were performed by 10 subjects, where each subject performed 10 times for each digit sign.

OUHANDS Dataset. The OUHANDS dataset comprises 10 signs acquired by the Intel RealSense F200 sensor [33]. This dataset includes hand and non-hand samples. Only the hand samples were selected for the present experiment. The hand samples were performed by 23 subjects, and 2150 and 1000 samples were adopted as the training and testing data, respectively.

4.2. Hand Detection and Depth Map Enhancement

The hand is assumed as the closest object to the camera, which is reasonable in practice. We detected the hand and enhanced its corresponding depth map to suppress the noise as well as improve the representation of the hand gesture. The first step applied Otsu's method [34] to select a threshold from the depth image. Pixel values smaller than the threshold were assumed as the background and set to zero. The second step applied the connected-component labeling algorithm [35] to group the non-zero pixels as foreground objects. In the NTU dataset, the objects closest to the camera were sometimes the knee regions of subjects seated on a chair. We thus selected the top foreground object as the hand because the hands are usually above the knees. The third step linearly scaled the pixel values in the hand region to 0–1 to enhance the hand texture.

When cropping the hands, the unequal width–height ratios hindered the batch learning because the image samples in a mini-batch should have identical width–height ratios. If the width was greater than the height, we resized the width to 64 pixels and maintained a constant width–height ratio. The height was expanded to 64 pixels by zero-padding; otherwise, we resized the height and expanded the width. Resizing and zero-padding did not alter the shape of the hand gesture.

4.3. Training and Testing

RFaNet was trained with a 0.9 momentum over 10 epochs and a 10^{-4} weight decay. The initial learning rate was 0.1, which was halved every 10 epochs. The proposed model was trained with a mini-batch size of 64 on an NVIDIA GeForce GTX 1080 Ti GPU using the PyTorch library.

The testing phase was implemented by leave-one-subject-out cross-validation (LOOCV). One subject was adopted as the testing data while the remaining subjects were adopted as the training data. The LOOCV was iterated until each subject was removed once. The LOOCV revealed whether RFaNet could be generalized to an unseen subject and whether RFaNet was robust to inter-subject variability, a common problem in practice.

We evaluated RFaNet in terms of classification accuracy. Furthermore, we computed the precision, recall, and F-score in comparison with state-of-the-art methods. We computed these measures as follows:

$$Accuracy = \frac{TP + TN}{TP + TN + FP + FN}, \tag{6}$$

$$Precision = \frac{TP}{TP + FP}, \tag{7}$$

$$Recall = \frac{TP}{TP + FN}, \tag{8}$$

$$F = 2\frac{Precision \cdot Recall}{Precision + Recall},\qquad(9)$$

where TP, TN, FP, and FN represent the numbers of true positives, true negatives, false positives, and false negatives, respectively. The F-score represents the harmonic mean of precision and recall.

4.4. Comparison of Different RF Blocks

To assess the effectiveness of the proposed NLRF block, we inserted NLRF and other blocks of receptive fields, namely ASPP and RFB, into the proposed RFaNet and compared their performances. These blocks processed multi-scale inputs. Table 1 presents the performance of RFaNet using different RF blocks. Note that only the NLRF block in RFaNet was replaced with ASPP or RFB. The NLRF block achieved a significant performance boost on both ASL and NTU datasets compared with ASPP and RFB.

Table 1. Performance comparison of RFaNet with different blocks of receptive fields, evaluated on the ASL and NTU datasets. We simply replaced the NLRF block with ASPP and RFB to evaluate the effects of these blocks. Numbers in parentheses indicate the standard deviation. #FLOPs: number of floating-point operations; #Param: number of parameters in the model. The symbol + indicates that the block was inserted into the proposed RFaNet. Bold values indicate the highest classification accuracy among the three blocks.

Block	#FLOPs (B)	#Param (M)	ASL (%)	NTU (%)
+ASPP	1.67	4.73	94.48(1.91)	95.80(4.61)
+RFB	3.19	10.69	94.60(1.90)	95.50(4.55)
+NLRF	3.06	5.45	**95.20(2.08)**	**96.50(3.63)**

4.5. Effect of Different Receptive Fields in NLRF Block

To examine the effect of varying the RFs in the NLRF block of RFaNet, Table 2 presents the performance of various configurations of the NLRF block on both ASL and NTU datasets. Each row indicates one configuration combining different branches of atrous convolution. Configurations 2–4 applied the vNL block to building non-local (long-range) connections across different branches of atrous convolution. The comparison of Configurations 1 and 2 shows that the vNL block improved the classification accuracy on both the ASL (+0.77%) and NTU (+0.10%) datasets. However, the computational cost increased in terms of the number of FLOPs (+2.80 B) and parameters (+1.96 M) due to the use of two vNL blocks. Notably, the vNL block was not applied to branches $r = 1$ and 3 because the RFs of these branches have a large overlap.

Table 2. Effects of various RFs in the NLRF block. The configuration $r = 1$ denotes the branch of atrous convolution with rate 1, and ✓ denotes that the branch was applied. The output feature maps of the branches with symbols * and † were further processed by the vNL block for capturing long-range dependencies. Configuration 3 adopted two vNL blocks to process branches $r = 1$ and 5 and branches $r = 1$ and 7, denoted by * and †, respectively. Numbers in parentheses indicate the standard deviation. A graphical illustration of the NLRF configuration is depicted in Figure 3c. Bold values indicate the highest classification accuracy among the four configurations.

	Configuration				#FLOPs	#Params	ASL	NTU
	$r=1$	$r=3$	$r=5$	$r=7$	vNL	(B)	(M)	(%)	(%)
1	✓	✓	✓	✓		1.67	4.73	94.48(1.91)	96.20(3.08)
2	✓*†	✓	✓*	✓†	✓	4.47	6.69	95.25(1.80)	96.30(3.43)
3	✓*	✓	✓*		✓	3.06	5.45	95.11(1.72)	96.25(3.70)
4	✓*	✓		✓*	✓	3.06	5.45	**95.20(2.08)**	**97.00(3.09)**

The comparison of configurations 2 and 3 shows that the classification accuracies slightly decreased on the ASL (−0.14%) and NTU (−0.05%) datasets when removing one vNL block. However, these classification accuracies were better than that of Configuration 1, which did not apply the vNL block. The comparison of Configurations 3 and 4 shows that connecting the branches $r = 1$ and 7 led to better performance than connecting the branches $r = 1$ and 5. Configuration 4 achieved comparable performance to Configuration 2 but saved computational cost. We selected configuration 4 as the NLRF configuration due to the tradeoff between accuracy and computational cost.

4.6. Qualitative Analysis of Various Receptive Fields in NLRF Block

Next, we analyzed the effect of changing the RFs in the NLRF block of RFaNet. Visual explanations were generated from the NLRF block using gradient-weighted class activation mapping (Grad-CAM) [36]. Grad-CAM can produce localization maps, which highlighted essential regions of the fingerspelling images corresponding to any decision of interest. Therefore, the discriminative features learned by NLRF could be visualized by Grad-CAM. Figure 6 shows the outcomes of each branch of atrous convolution from the NLRF block. The atrous convolution with a large rate captured the long-range dependency, whereas that with a small rate captured neighboring dependency. As the rate increased, the grid effect was observed in the branch of atrous convolution with rates 5 and 7. The localization maps highlighted the essential regions with checkerboard patterns, losing some neighboring information because the regions between two pixels of the convolutional kernel were not considered. Similar results from atrous convolutions with large rates were reported in [29].

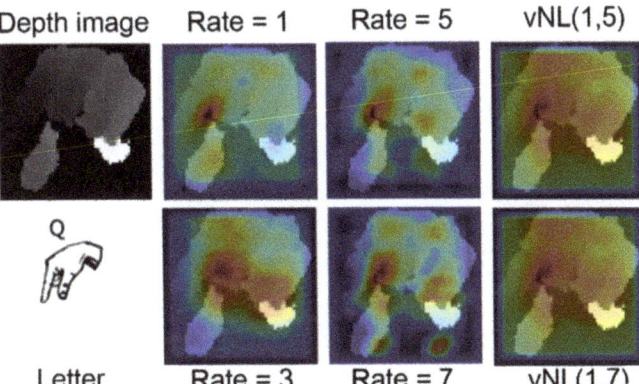

Figure 6. Visualization of outcomes of each branch from the NLRF block. The left-most column shows the depth image and its corresponding label. The middle two columns show the outcome of each branch of atrous convolution. vNL(1,5) and vNL(1,7) represent the outcomes of the vNL block determining the dependencies between the branches of atrous convolutions with rates 1 and 5 and rates 1 and 7, respectively.

The rightmost column of Figure 6 shows that the vNL block integrated the atrous convolution with small and large rates. However, the outcomes of the two vNL blocks NL(1,5) and NL(1,7) were similar (the notation is explained in the caption of Figure 6), which suggested that integrating the atrous convolutions with rates 1 and 5 and rates 1 and 7 gives similar contributions. Therefore, we maintained one of the vNL blocks and selected NL(1,7), i.e., Configuration 4 in Table 2, because it provides slightly higher performance and longer-range dependency than NL(1,5).

4.7. Effects of DFA and NLRF Blocks

To assess the effectiveness of the DFA and NLRF blocks, we performed comprehensive ablation experiments on the ASL and NTU datasets. Table 3 shows the performances of RFaNet in various configurations. Here, a VGG-13 was adopted as the backbone. The comparison of rows 1 and 2 presents a performance boost on both datasets when adopting the DFA block, which demonstrates that selecting the representative finger regions from the depth image facilitated accuracy improvement. The configuration of row 3 adopted a VGG-9 as backbone instead of a VGG-13 because the insertion of the NLRF block increased three convolutional layers and one vNL block. Therefore, the NLRF block was inserted into the VGG-9 for a fair comparison. Inserting the NLRF block significantly improved the performance on both datasets (cf. rows 1 and 3 of Table 3), demonstrating the effectiveness of building short- and long-range dependencies. Moreover, employing both the DFA and NLRF blocks significantly improved the accuracy for the ASL (+1.7%) and NTU (+7.0%) datasets (cf. rows 1 and 4 of Table 3). For the computational cost, the number of parameters of the model is less than that of the backbone (−5.14 M).

Table 3. Ablation study for various configurations of RFaNet on the ASL and NTU datasets in terms of accuracy (%). Notably, the VGG-13 has four more convolutional layers than the VGG-9. These four convolutional layers were inserted before the global average pooling layer. Numbers in parentheses indicate the standard deviation. #FLOPs: number of floating-point operations; #Param: number of parameters of a model. Bold values indicate the highest classification accuracy among the four configurations.

	Configuration			#FLOPs (B)	#Param (M)	ASL (%)	NTU (%)
	Backbone	DFA	NLRF				
1	VGG-13			1.43	10.59	93.50(2.30)	89.50(6.59)
2	VGG-13	✓		1.59	10.63	94.26(1.94)	91.40(4.72)
3	VGG-9		✓	2.91	5.41	94.87(2.33)	94.90(4.86)
4	VGG-9	✓	✓	3.06	5.45	**95.20(2.08)**	**96.50(3.63)**

4.8. Qualitative Analysis of DFA and NLRF Blocks

We conducted a qualitative analysis of the DFA and NLRF blocks in RFaNet. Figure 7 shows the effects of the DFA and NLRF blocks on the ASL dataset. As shown in the left two columns, RFaNet without the DFA block highlighted only the hand contours. By contrast, RFaNet with the DFA block highlighted the fingers in the depth image while ignoring the wrist, which was irrelevant to the fingerspelling sign. Furthermore, using the DFA block increased the softmax score of the ground-truth class, leading to correct classification.

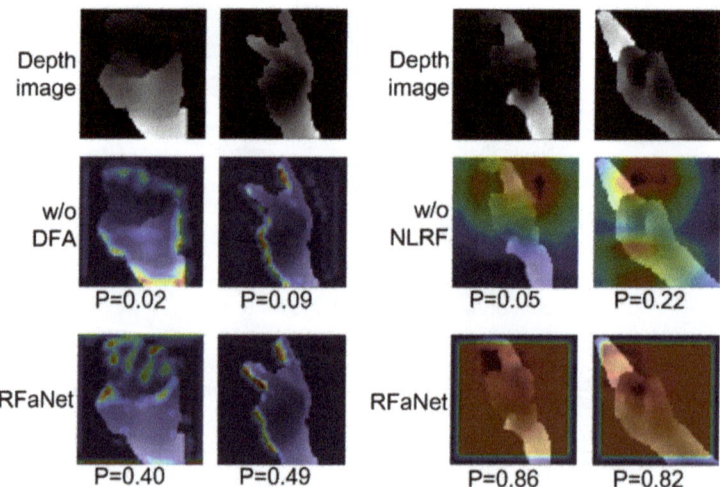

Figure 7. Grad-CAM visualization of the effects of the DFA and NLRF blocks on the ASL dataset. The left two columns show the feature maps of the first bottleneck layer of RFaNet to analyze the effect of the DFA block. The right two columns show the feature maps of the last bottleneck layer of RFaNet to analyze the effect of the NLRF block. w/o: without; P: softmax score of the ground-truth class.

The right two columns of Figure 7 show that RFaNet without the NLRF block emphasized the background rather than the fingers, leading to a low softmax score of the ground-truth class. When the NLRF block was inserted, RFaNet exploited the neighboring and long-range dependencies to emphasize the key fingers related to the fingerspelling sign. Therefore, fingerspelling signs in which the key fingers cover a large area (column 3) and wide posture variation (column 4) were correctly classified.

4.9. Comparison with State-of-The-Art Methods

We compared the performances of RFaNet and state-of-the-art methods on the ASL and NTU datasets. Table 4 lists the classification accuracy. For the ASL dataset, RFaNet outperformed the state-of-the-art methods (95.30%). For the NTU dataset, RFaNet did not outperform the state-of-the-art methods (98.00%). However, the state-of-the-art methods did not simultaneously achieve the highest accuracy on both datasets, and RFaNet was competitive against all compared methods. The high accuracies of RFaNet on both datasets demonstrated the generalization ability of RFaNet for various fingerspelling tasks.

Table 4. Comparisons with state-of-the-art methods using LOOCV evaluation on the ASL and NTU datasets. The "Method" column shows the classifiers used in state-of-the-art methods, where * indicates that the feature descriptor and classifier are jointly trained in the method. Mod: modality; A: accuracy; P: precision; R: recall; F: F-score. Bold values indicate the highest classification accuracy among the state-of-the-art methods.

			ASL				NTU
Study	**Method**	**Mod.**	A (%)	P (%)	R (%)	F (%)	A (%)
Pugeault et al. [3]	RF	D	49.00	–	–	–	–
Kuznetsova et al. [37]	RF	D	57.00	–	–	–	–
Wang et al. [38]	SVM	D	58.30	–	–	–	91.10
Dong et al. [39]	RF	RGBD	70.00	–	–	–	–
Kane et al. [40]	SVM	D	71.58	–	–	–	90.75
Wang et al. [17]	TM	RGBD	75.80	–	–	–	99.60
Suau et al. [41]	RF	RGBD	76.10	–	–	–	–
Feng et al. [42]	SVM	D	78.70	–	–	–	**100**
Warchoł et al. [43]	HMM	D	78.80	–	–	–	–
Ameen et al. [44]	CNN *	RGBD	80.34	82.00	80.00	79.20	–
Nai et al. [45]	RF	D	81.10	–	–	–	–
Maqueda et al. [46]	SVM	RGB	83.70	–	–	–	95.90
Zhang et al. [16]	SVM	D	83.80	–	–	–	94.50
Keskin et al. [47]	SCF	D	84.30	–	–	–	–
Rady et al. [48]	CNN *	RGBD	84.67	–	–	–	99.85
Aly et al. [49]	SVM	D	88.70	–	–	–	–
Rakowski et al. [50]	ResNet *	RGBD	90.60	91.80	90.60	90.30	–
Tao et al. [18]	CNN *	D	92.70	93.50	92.40	91.71	**100**
Yang et al. [27]	DDaNet *	RGBD	93.53	94.10	93.48	93.26	96.10
Ours	RFaNet	D	95.30	95.32	95.70	95.51	98.00

5. Extensive Experimental Results of RFaNet in Transfer Learning

The data annotation and collection of fingerspelling requires specialized domain knowledge and expert interpreters. Thus, large-scale datasets are not commonly available for fingerspelling recognition. Transferring the representation of hand gestures from a large- to a small-scale dataset is always in demand. We evaluated the effectiveness of RFaNet in transferring knowledge from the large-scale ASL dataset to the small-scale NTU and OUHANDS datasets. ASL, NTU, and OUHANDS datasets are commonly used fingerspelling datasets and comprise 60,000, 1000, and 3000 labeled samples, respectively. These three datasets share similar hand gestures, even when they belong to different labels, as shown in Figure 5.

5.1. Implementation Details of Transfer Learning

Transfer learning for fingerspelling recognition was implemented by the following process. First, RFaNet was pre-trained with the ASL dataset (source domain). Second, the last fully connected layers and the corresponding softmax layer were replaced according to the number of classes in the target dataset. Third, the initial two bottleneck layers had their parameters frozen (shared with the source domain) when considering the OUHANDS dataset as the target dataset. The first three bottleneck layers had their parameters frozen when considering the NTU dataset as the target dataset. The number of frozen bottleneck layers differed for the OUHANDS and NTU datasets because the OUHANDS dataset contains more training data (3150) and is larger than the NTU dataset (1000). If the number of parameters requiring fine-tuning and the target dataset were small, the model would result in overfitting [51]. In the fourth step, the remaining model parameters were fine-tuned on the target dataset.

5.2. Quantitative Results of Transfer Learning on NTU Dataset

Table 5 shows the experimental results of transfer learning when considering the NTU dataset as the target dataset. For comparison, we implemented transfer learning on DDaNet [27], a state-of-the-art method for the ASL dataset that adopts the color (RGB) and depth modalities as inputs. The transfer learning protocol for DDaNet was identical to that of RFaNet. Applying the transfer learning to RFaNet improved the accuracy compared with RFaNet without transfer learning (+1.00%). Furthermore, the number of parameters for RFaNet was less than that of DDaNet (−16.37 M), making mobile applications feasible.

Table 5. Results on the transfer learning where the NTU dataset is the target dataset. Numbers in parentheses indicate the standard deviation of the LOOCV across 10 subjects. Mod.: modality; #Param: number of parameters of the model; w/TF: with transfer learning; w/o TF: without transfer learning. Bold values indicate the highest classification accuracy among the four methods.

Study	Method	Mod.	#Param	Accuracy (%)
Yang et al. [27]	DDaNet w/TF	RGBD	21.24 M	96.10 (4.12)
Yang et al. [27]	DDaNet w/o TF	RGBD	21.24 M	87.90 (4.75)
Ours	RFaNet w/TF	D	5.45 M	97.00 (3.09)
Ours	RFaNet w/o TF	D	5.45 M	**98.00 (1.56)**

5.3. Quantitative Results of Transfer Learning on OUHANDS Dataset

Table 6 shows the experimental results of transfer learning when considering the OUHANDS dataset as the target dataset. After transfer learning, the accuracy and F-score of DDaNet were lower than those of DDaNet without transfer learning (−0.80% and −0.85%, respectively). However, after transfer learning, RFaNet showed improved accuracy and F-score compared with RFaNet without transfer learning (+2.60% and +2.66%, respectively). Furthermore, RFaNet outperformed the state-of-the-art methods in terms of accuracy and F-score (92.90% and 93.00%, respectively), demonstrating the benefits of learning the representations of hand gestures using depth modality from a large-scale dataset (the ASL dataset).

5.4. Qualitative Results of Transfer Learning

For a qualitative analysis of transfer learning by RFaNet, we generated localization maps using Grad-CAM [36] to highlight the essential regions corresponding to any decisions of interest. This analysis visualized the representation of the hand gestures learned by RFaNet during transfer learning. Figure 8 shows the qualitative analysis of transfer learning where NTU and OUHANDS datasets are the target datasets. The localization maps of the NLRF layer revealed that RFaNet without transfer learning emphasized the regions in the background, as shown in the number "6" of NTU and the letter "c" of OUHANDS. Although the ring finger and thumb, respectively, were highlighted in letters "f" and "k" of OUHANDS, the other key fingers of these hand gestures were not emphasized, leading to a low softmax score of the ground-truth class. After transferring the representation of the hand gestures learned from the ASL dataset, the key fingers of the hand gestures were highlighted, and the softmax score of the ground-truth class was increased, as shown in the third row of Figure 8.

Table 6. Results of the transfer learning where OUHANDS is the target dataset. Numbers in parentheses indicate the performance difference between networks with and without transfer learning. Mod.: modality; #Param: number of parameters of the model; w/TF: with transfer learning; w/o TF: without transfer learning. A: accuracy; F: F-score; HGR-Net: hand gesture recognition network; HOG: histogram of oriented gradients; SVM: support vector machine. Bold values indicate the highest classification accuracy among all methods.

Study	Method	Mod.	#Param	A	F
He et al. [52]	ResNet-50	RGB	23.60 M	–	81.30
Huang et al. [53]	DenseNet-121	RGB	7.04 M	–	82.80
Howard et al. [54]	MobileNet	RGB	3.22 M	–	86.50
Dadashzadeh et al. [55]	HGR-Net	RGB	0.499 M	–	88.10
Matilainen et al. [33]	HOG+SVM	RGB	–	83.25	–
Yang et al. [27]	DDaNet w/TF	RGBD	21.24 M	88.90	89.10
Yang et al. [27]	DDaNet w/o TF	RGBD	21.24 M	88.10 (−0.80)	88.25 (−0.85)
Ours	RFaNet w/TF	D	5.45 M	90.30	90.34
Ours	RFaNet w/o TF	D	5.45 M	**92.90** (+2.60)	**93.00** (+2.66)

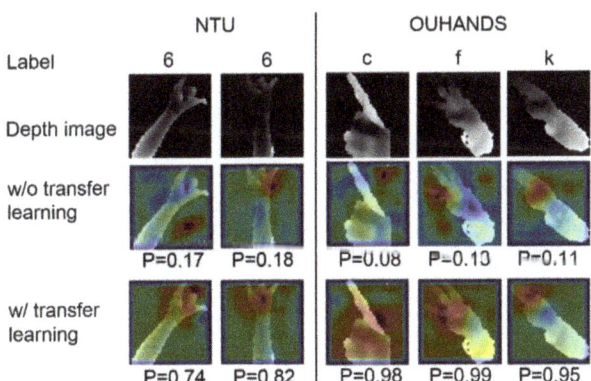

Figure 8. Qualitative analysis of transfer learning on the NTU and OUHANDS datasets where "w/o" and "w/" represent without and with, respectively. The NLRF layer was visualized by Grad-CAM. Transfer learning was implemented by pre-training RFaNet with the ASL dataset and fine-tuning it with the OUHANDS dataset. Because the amount of data in the NTU dataset is 1/3 of that of the OUHANDS dataset, two and three samples were provided for the NTU and OUHANDS datasets, respectively. P is the softmax score of the ground-truth class.

5.5. Network Visualization of Transfer Learning

In addition to the NLRF layer, we qualitatively visualized the output of each bottleneck layer to demonstrate the effectiveness of RFaNet during transfer learning. Figure 9 shows the outcomes from the initial three bottleneck layers for three examples. When RFaNet learned the representation of the hand gestures from the ASL dataset, it could more efficiently extract the low-level features in the small-scale target dataset than it could without transfer learning. The key fingers were then accurately localized, leading to correct classification. This result agreed with the empirical evidence showing that the initial bottleneck layers learned the low-level features that could be shared across different tasks [56].

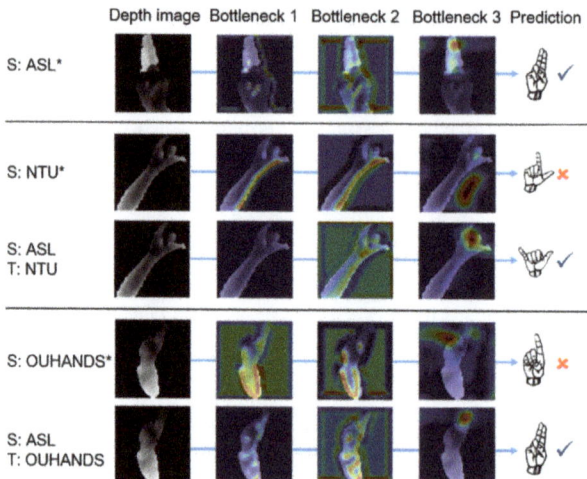

Figure 9. Visualization of the learned features when applying transfer learning to RFaNet, where S and T represent the source and target dataset, respectively. Each row indicates the feature maps of three bottlenecks of RFaNet. The asterisk indicates that RFaNet was trained using only the source data and evaluated on the source data. The third and fifth rows indicate that RFaNet was pretrained with the source data, fine-tuned with the target data, and evaluated on the target data. This study considered the ASL dataset as the source data due to the sufficiently large training data and considered the OUHANDS and NTU datasets as the target data due to the relatively small amount of training data. The icons in the prediction column were reproduced from [57].

5.6. Failure Modes

Figure 10 shows some failure modes of RFaNet on the ASL and NTU datasets. Our model failed to capture the neighboring and inter-region dependencies of widely variable hand postures. When the fingers extended outside the palm region, they were not correctly highlighted in the localization maps, leading to incorrect classification. Dealing with large hand-posture variations is left for future work.

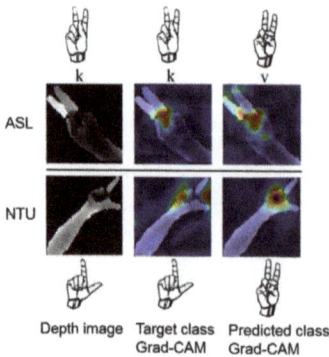

Figure 10. Failure modes of RFaNet on the ASL and NTU datasets. The first and fourth rows represent the labels corresponding to the images and feature maps, respectively. The second and third rows represent the depth images and their Grad-CAM visualizations, respectively, according to their target and predicted classes.

6. Discussion

6.1. Effectiveness of the DFA and NLRF Blocks

The proposed DFA and NLRF blocks are inserted at the top and bottom of RFaNet, respectively. The DFA block highlights the fingers in the depth image. The NLRF block increases the size of the receptive fields and builds long-range connections across the finger features, thus facilitating fingerspelling recognition. This result demonstrates that building long-range connections across the branches of atrous convolutions with rates $r = 1$ and $r = 7$ facilitates the network's learning of discriminative features related to fingerspelling. Furthermore, integrating the small and large receptive fields by the vNL blocks improves fingerspelling recognition. The vNL block integrates the short-range and long-range dependencies and exploits the relation between local and non-local interactions. The NLRF block could effectively capture the fine fingerspelling details and important features across the fingers. Furthermore, this integration allowed RFaNet to recognize letter signs whose important fingers possess long-range dependency and hand shapes with high inter-class similarity. The DFA and NLRF blocks highlight the finger regions and explore the fingers' dependencies, contributing to the performance boost of RFaNet.

6.2. Transfer Learning for Fingerspelling Recognition

The recognition accuracy of RFaNet on small-scale datasets (e.g., the NTU and OUHANDS datasets) can be improved by transferring the representations of hand gestures learned from large-scale datasets (e.g., the ASL dataset). The above experimental results show that the proposed RFaNet learned better representations of the hand gestures from a large-scale dataset than did DDaNet. As DDaNet learns the representation from both color and depth modalities, it may learn to highlight the background information revealed in the color modality corresponding to any decision of interest. This learning can degrade the transfer learning because the low-level features relevant to the background differ across datasets. Therefore, the initial bottleneck layers with frozen parameters may not be shared across the source and target domains. However, RFaNet learns only from the depth modality. As the DFA block of RFaNet facilitates the separation of hand gestures from the background and highlights the fingers, the low-level features hardly involve the background information. Therefore, the initial bottleneck layers pre-trained on the source domain improved the classification accuracy in the target domain. This result demonstrates that during transfer learning, RFaNet can boost fingerspelling recognition on small-scale datasets without the effect of complex background information.

RFaNet efficiently learned the representations of hand gestures from a large-scale dataset and facilitated the learning of a small-scale target dataset. The reasons are explained here. First, as RFaNet processes only depth images, the hand gestures are not easily affected by the complex background. Therefore, RFaNet can effectively transfer the representation of the hand gesture learned from the ASL dataset to the small-scale NTU/OUHANDS datasets, leading to improved recognition performance. Second, the DFA block in the most initial layer emphasizes the fingers and palm regions, indicating that the learning of hand-gesture representations is unaffected by gesture-irrelevant factors. Therefore, RFaNet facilitated transfer learning when the training data of the target domain were insufficient.

6.3. Implementation in Actual Application

The implementation of the proposed fingerspelling recognition system in actual experiments consists of two factors: hardware and software. The hardware factor considers the depth camera and experimental environment. The training datasets were collected by a Microsoft Kinect sensor (ASL and NTU) and a RealSense F200 sensor (OUHANDS). Both depth cameras acquire depth images with a depth resolution of 1 mm and a spatial resolution of 640×480 pixels. The distance from the subject to the depth camera is in a range of 230–800 mm in an indoor environment. The software factor considers hand detection and depth map enhancement. We detected the hand and enhanced its corresponding depth map to suppress the noise as well as improve the representation of the hand gesture.

When using a new depth camera, the depth image should possess a depth resolution of 1 mm and a spatial resolution of 640 × 480 pixels. Furthermore, the subject is kept at a distance in a range of 230–800 mm from the depth sensor in order to obtain a hand image with quality similar to that of the training datasets. If the hardware meets these requirements in an indoor environment, the proposed fingerspelling recognition system could be implemented using a new depth camera in actual experiments.

7. Conclusions

We proposed and evaluated RFaNet, a network that highlights the finger regions and builds inter-finger relations for fingerspelling recognition. RFaNet aggregates the low-level features in hand depth and non-forearm images to focus on the fingers. It fuses the high-level multi-scale features of various RFs to model the neighboring and inter-region dependencies between fingers, which makes the sign representation invariant to the viewpoint and thus reduces the intra-class variability. In experimental evaluations on the ASL dataset, RFaNet outperformed current state-of-the-art methods. When applied to a small-scale fingerspelling dataset with insufficiently labeled data, RFaNet leverages the depth representations learned from a large-scale dataset to boost the fingerspelling recognition on the small-scale dataset. Using only depth images in RFaNet facilitated transfer learning on limited training datasets without requiring expensive fingerspelling annotations. This technique can improve communication between deaf and hearing people. Large hand posture variations may affect neighboring and inter-region dependencies. Therefore, the question of how to build inter-finger relations under large hand posture variations is left for future work.

Author Contributions: Conceptualization, S.-H.Y., Y.-M.C. and Y.-P.C.; methodology, S.-H.Y. and Y.-M.C.; software, Y.-M.C. and J.-W.H.; validation, Y.-M.C. and J.-W.H.; formal analysis, S.-H.Y. and Y.-M.C.; investigation, S.-H.Y., Y.-M.C., and J.-W.H.; data curation, Y.-M.C. and J.-W.H.; writing—original draft preparation, S.-H.Y. and Y.-M.C.; writing—review and editing, S.-H.Y. and Y.-M.C.; visualization, Y.-M.C. and S.-H.Y.; supervision, S.-H.Y. and Y.-P.C.; project administration, S.-H.Y. and Y.-P.C.; funding acquisition, S.-H.Y. and Y.-P.C. All authors have read and agreed to the published version of the manuscript.

Funding: This work was supported in part by the Ministry of Science and Technology of Taiwan MOST 108-2221-E-009-119, 109-2221-E-009-049, 110-2221-E-A49-122, 110-2636-E-006-021 (Young Scholar Fellowship Program); in part by the Headquarters of University Advancement at National Cheng Kung University, Ministry of Education; in part by National Cheng Kung University Hospital, Taiwan.

Data Availability Statement: Publicly available datasets were analyzed in this study. This data can be found here: https://www.kaggle.com/mrgeislinger/asl-rgb-depth-fingerspelling-spelling-it-out for ASL; http://eeeweba.ntu.edu.sg/computervision/people/home/renzhou/HandGesture.htm for NTU; and https://www.kaggle.com/mumuheu/ouhands for OUHANDS.

Acknowledgments: The authors would like to thank Pugeault and Bowden, Ren et al., and Matilainen et al., respectively, for making the ASL Fingerspelling Dataset, NTU Digit Dataset, and OUHANDS Dataset publicly available.

Conflicts of Interest: The authors declare no conflict of interest.

References

1. Padden, C.A.; Gunsauls, D.C. How the alphabet came to be used in a sign language. *Sign Lang. Stud.* **2003**, *4*, 10–33. [CrossRef]
2. Tsai, Y.-S.; Hsu, L.-H.; Hsieh, Y.-Z.; Lin, S.-S. The real-time depth estimation for an occluded person based on a single image and OpenPose method. *Mathematics* **2020**, *8*, 1333. [CrossRef]
3. Pugeault, N.; Bowden, R. Spelling it out: Real-time ASL fingerspelling recognition. In Proceedings of the IEEE International Conference on Computer Vision Workshops (ICCV Workshops), Barcelona, Spain, 6–13 November 2011; pp. 1114–1119.

4. Rioux-Maldague, L.; Giguère, P. Sign language fingerspelling classification from depth and color images using a deep belief network. In Proceedings of the IEEE Canadian Conference on Computer and Robot Vision, Montreal, QC, Canada, 6–9 May 2014; pp. 92–97.
5. Tian, M.; Yi, S.; Li, H.; Li, S.; Zhang, X.; Shi, J.; Yan, J.; Wang, X. Eliminating background-bias for robust person re-identification. In Proceedings of the IEEE Conference on Computer Vision and Pattern Recognition, Salt Lake City, UT, USA, 18–23 June 2018; pp. 5794–5803.
6. Modanwal, G.; Sarawadekar, K. A robust wrist point detection algorithm using geometric features. *Pattern Recognit. Lett.* **2018**, *110*, 72–78. [CrossRef]
7. Chen, L.-C.; Papandreou, G.; Kokkinos, I.; Murphy, K.; Yuille, A.L. Deeplab: Semantic image segmentation with deep convolutional nets, atrous convolution, and fully connected crfs. *IEEE Trans. Pattern Anal. Mach. Intell.* **2017**, *40*, 834–848. [CrossRef]
8. Liu, S.; Huang, D. Receptive field block net for accurate and fast object detection. In Proceedings of the European Conference on Computer Vision (ECCV), Munich, Germany, 8–14 September 2018; pp. 385–400.
9. Wang, X.; Girshick, R.; Gupta, A.; He, K. Non-local neural networks. In Proceedings of the IEEE Conference on Computer Vision and Pattern Recognition, Salt Lake City, UT, USA, 18–23 June 2018; pp. 7794–7803.
10. Cui, Y.; Song, Y.; Sun, C.; Howard, A.; Belongie, S. Large scale fine-grained categorization and domain-specific transfer learning. In Proceedings of the IEEE Conference on Computer Vision and Pattern Recognition, Salt Lake City, UT, USA, 18–23 June 2018; pp. 4109–4118.
11. Nihal, R.A.; Rahman, S.; Broti, N.M.; Deowan, S.A. Bangla Sign Alphabet Recognition with Zero-shot and Transfer Learning. *Pattern Recognit. Lett.* **2021**, *150*, 84–93. [CrossRef]
12. Deng, J.; Dong, W.; Socher, R.; Li, L.-J.; Li, K.; Fei-Fei, L. Imagenet: A large-scale hierarchical image database. In Proceedings of the IEEE Conference on Computer Vision and Pattern Recognition, Miami, FL, USA, 20–25 June 2009; pp. 248–255.
13. Bird, J.J.; Ekárt, A.; Faria, D.R. British sign language recognition via late fusion of computer vision and leap motion with transfer learning to american sign language. *Sensors* **2020**, *20*, 5151. [CrossRef]
14. Simonyan, K.; Zisserman, A. Very deep convolutional networks for large-scale image recognition. *arXiv* **2014**, arXiv:1409.1556.
15. Hu, Y.; Zhao, H.-F.; Wang, Z.-G. Sign language fingerspelling recognition using depth information and deep belief networks. *Int. J. Pattern Recognit. Artif. Intell.* **2018**, *32*, 1850018. [CrossRef]
16. Zhang, C.; Tian, Y. Histogram of 3D facets: A depth descriptor for human action and hand gesture recognition. *Comput. Vis. Image Underst.* **2015**, *139*, 29–39. [CrossRef]
17. Wang, C.; Liu, Z.; Chan, S.-C. Superpixel-based hand gesture recognition with kinect depth camera. *IEEE Trans. Multimed.* **2014**, *17*, 29–39. [CrossRef]
18. Tao, W.; Leu, M.C.; Yin, Z. American Sign Language alphabet recognition using Convolutional Neural Networks with multiview augmentation and inference fusion. *Eng. Appl. Artif. Intell.* **2018**, *76*, 202–213. [CrossRef]
19. Wang, Y.; Wu, T.; Yang, J.; Wang, L.; An, W.; Guo, Y. DeOccNet: Learning to see through foreground occlusions in light fields. In Proceedings of the IEEE/CVF Winter Conference on Applications of Computer Vision, Village, CO, USA, 1–5 March 2020; pp. 118–127.
20. Tan, Y.S.; Lim, K.M.; Tee, C.; Lee, C.P.; Low, C.Y. Convolutional neural network with spatial pyramid pooling for hand gesture recognition. *Neural Comput. Appl.* **2021**, *33*, 5339–5351. [CrossRef]
21. Lu, R.; Xue, F.; Zhou, M.; Ming, A.; Zhou, Y. Occlusion-shared and feature separated network for occlusion relationship reasoning. In Proceedings of the IEEE/CVF International Conference on Computer Vision, Seoul, Korea, 27–28 October 2019; pp. 10343–10352.
22. Hu, J.; Shen, L.; Sun, G. Squeeze-and-excitation networks. In Proceedings of the IEEE Conference on Computer Vision and Pattern Recognition, Salt Lake City, UT, USA, 18–22 June 2018; pp. 7132–7141.
23. Vaswani, A.; Shazeer, N.; Parmar, N.; Uszkoreit, J.; Jones, L.; Gomez, A.N.; Kaiser, Ł.; Polosukhin, I. Attention is all you need. In Proceedings of the Advances in Neural Information Processing Systems, Long Beach, CA, USA, 4–9 December 2017; pp. 5998–6008.
24. Wang, L.; Wang, Y.; Liang, Z.; Lin, Z.; Yang, J.; An, W.; Guo, Y. Learning parallax attention for stereo image super-resolution. In Proceedings of the IEEE Conference on Computer Vision and Pattern Recognition, Long Beach, CA, USA, 16–20 June 2019; pp. 12250–12259.
25. Han, B.; Yin, J.; Luo, X.; Jia, X. Multibranch Spatial-Channel Attention for Semantic Labeling of Very High-Resolution Remote Sensing Images. *IEEE Geosci. Remote. Sens. Lett.* **2020**, 1–5. [CrossRef]
26. Liu, N.; Zhang, N.; Han, J. Learning selective self-mutual attention for RGB-D saliency detection. In Proceedings of the IEEE/CVF Conference on Computer Vision and Pattern Recognition, Seattle, WA, USA, 16–18 June 2020; pp. 13756–13765.
27. Yang, S.-H.; Chen, W.-R.; Huang, W.-J.; Chen, Y.-P. DDaNet: Dual-Path Depth-Aware Attention Network for Fingerspelling Recognition Using RGB-D Images. *IEEE Access* **2020**, *9*, 7306–7322. [CrossRef]
28. Nair, V.; Hinton, G.E. Rectified linear units improve restricted boltzmann machines. In Proceedings of the 27th International Conference on International Conference on Machine Learning, Haifa, Israel, 21–24 June 2010; pp. 807–814.
29. Wang, P.; Chen, P.; Yuan, Y.; Liu, D.; Huang, Z.; Hou, X.; Cottrell, G. Understanding convolution for semantic segmentation. In Proceedings of the 2018 IEEE Winter Conference on Applications of Computer Vision (WACV), Lake Tahoe, NV, USA, 12–15 March 2018; pp. 1451–1460.

30. Cao, Y.; Xu, J.; Lin, S.; Wei, F.; Hu, H. Gcnet: Non-local networks meet squeeze-excitation networks and beyond. In Proceedings of the IEEE/CVF International Conference on Computer Vision Workshops, Seoul, Korea, 27–28 October 2019; pp. 1–10.
31. Liu, Z.; Li, J.; Shen, Z.; Huang, G.; Yan, S.; Zhang, C. Learning efficient convolutional networks through network slimming. In Proceedings of the IEEE International Conference on Computer Vision, Venice, Italy, 22–29 October 2017; pp. 2736–2744.
32. Ren, Z.; Yuan, J.; Zhang, Z. Robust hand gesture recognition based on finger-earth mover's distance with a commodity depth camera. In Proceedings of the 19th ACM International Conference on Multimedia, Scottsdale, AZ, USA, 28 November–1 December 2011; pp. 1093–1096.
33. Matilainen, M.; Sangi, P.; Holappa, J.; Silvén, O. OUHANDS database for hand detection and pose recognition. In Proceedings of the Sixth International Conference on Image Processing Theory, Tools and Applications (IPTA), Oulu, Finland, 12–15 December 2016; pp. 1–5.
34. Otsu, N. A threshold selection method from gray-level histograms. *IEEE Trans. Syst. Man Cybern.* **1979**, *9*, 62–66. [CrossRef]
35. Rosenfeld, A.; Pfaltz, J.L. Sequential operations in digital picture processing. *JACM* **1966**, *13*, 471–494. [CrossRef]
36. Selvaraju, R.R.; Cogswell, M.; Das, A.; Vedantam, R.; Parikh, D.; Batra, D. Grad-cam: Visual explanations from deep networks via gradient-based localization. In Proceedings of the IEEE International Conference on Computer Vision, Venice, Italy, 22–29 October 2017; pp. 618–626.
37. Kuznetsova, A.; Leal-Taixé, L.; Rosenhahn, B. Real-time sign language recognition using a consumer depth camera. In Proceedings of the IEEE International Conference on Computer Vision Workshops, Sydney, Australia, 2–8 December 2013; pp. 83–90.
38. Wang, Y.; Yang, R. Real-time hand posture recognition based on hand dominant line using kinect. In Proceedings of the 2013 IEEE International Conference on Multimedia and Expo Workshops (ICMEW), San Jose, CA, USA, 15–19 July 2013; pp. 1–4.
39. Dong, C.; Leu, M.C.; Yin, Z. American sign language alphabet recognition using microsoft kinect. In Proceedings of the IEEE Conference on Computer Vision and Pattern Recognition Workshops, Boston, MA, USA, 7–12 June 2015; pp. 44–52.
40. Kane, L.; Khanna, P. A framework for live and cross platform fingerspelling recognition using modified shape matrix variants on depth silhouettes. *Comput. Vis. Image Underst.* **2015**, *141*, 138–151. [CrossRef]
41. Suau, X.; Alcoverro, M.; López-Méndez, A.; Ruiz-Hidalgo, J.; Casas, J.R. Real-time fingertip localization conditioned on hand gesture classification. *Image Vis. Comput.* **2014**, *32*, 522–532. [CrossRef]
42. Feng, B.; He, F.; Wang, X.; Wu, Y.; Wang, H.; Yi, S.; Liu, W. Depth-projection-map-based bag of contour fragments for robust hand gesture recognition. *IEEE Trans. Hum. Mach. Syst.* **2016**, *47*, 511–523. [CrossRef]
43. Warchoł, D.; Kapuściński, T.; Wysocki, M. Recognition of fingerspelling sequences in polish sign language using point clouds obtained from depth images. *Sensors* **2019**, *19*, 1078. [CrossRef]
44. Ameen, S.; Vadera, S. A convolutional neural network to classify American Sign Language fingerspelling from depth and colour images. *Expert Syst.* **2017**, *34*, e12197. [CrossRef]
45. Nai, W.; Liu, Y.; Rempel, D.; Wang, Y. Fast hand posture classification using depth features extracted from random line segments. *Pattern Recognit.* **2017**, *65*, 1–10. [CrossRef]
46. Maqueda, A.I.; del-Blanco, C.R.; Jaureguizar, F.; García, N. Human–computer interaction based on visual hand-gesture recognition using volumetric spatiograms of local binary patterns. *Comput. Vis. Image Underst.* **2015**, *141*, 126–137. [CrossRef]
47. Keskin, C.; Kıraç, F.; Kara, Y.E.; Akarun, L. Hand pose estimation and hand shape classification using multi-layered randomized decision forests. In Proceedings of the European Conference on Computer Vision, Florence, Italy, 7–13 October 2012; pp. 852–863.
48. Rady, M.A.; Youssef, S.M.; Fayed, S.F. Smart Gesture-based Control in Human Computer Interaction Applications for Special-need People. In Proceedings of the IEEE Novel Intelligent and Leading Emerging Sciences Conference (NILES), Cairo, Egypt, 28–30 October 2019; pp. 244–248.
49. Aly, W.; Aly, S.; Almotairi, S. User-Independent American Sign Language Alphabet Recognition Based on Depth Image and PCANet Features. *IEEE Access* **2019**, *7*, 123138–123150. [CrossRef]
50. Rakowski, A.; Wandzik, L. Hand shape recognition using very deep convolutional neural networks. In Proceedings of the 2018 International Conference on Control and Computer Vision, Singapore, 15–18 June 2018; pp. 8–12.
51. Yosinski, J.; Clune, J.; Bengio, Y.; Lipson, H. How transferable are features in deep neural networks? *Adv. Neural Inf. Process. Syst.* **2014**, *27*, 3320–3328.
52. He, K.; Zhang, X.; Ren, S.; Sun, J. Deep residual learning for image recognition. In Proceedings of the IEEE Conference on Computer Vision and Pattern Recognition, Las Vegas, NV, USA, 27–30 June 2016; pp. 770–778.
53. Huang, G.; Liu, Z.; Van Der Maaten, L.; Weinberger, K.Q. Densely connected convolutional networks. In Proceedings of the IEEE Conference on Computer Vision and Pattern Recognition, Honolulu, HI, USA, 21–26 July 2017; pp. 4700–4708.
54. Howard, A.G.; Zhu, M.; Chen, B.; Kalenichenko, D.; Wang, W.; Weyand, T.; Andreetto, M.; Adam, H. Mobilenets: Efficient convolutional neural networks for mobile vision applications. *arXiv* **2017**, arXiv:1704.04861.
55. Dadashzadeh, A.; Targhi, A.T.; Tahmasbi, M.; Mirmehdi, M. HGR-Net: A fusion network for hand gesture segmentation and recognition. *IET Comput. Vis.* **2019**, *13*, 700–707. [CrossRef]
56. Tajbakhsh, N.; Shin, J.Y.; Gurudu, S.R.; Hurst, R.T.; Kendall, C.B.; Gotway, M.B.; Liang, J. Convolutional neural networks for medical image analysis: Full training or fine tuning? *IEEE Trans. Med. Imaging* **2016**, *35*, 1299–1312. [CrossRef]
57. Wikipedia. American Manual Alphabet. Available online: http://en.wikipedia.org/wiki/American_manual_alphabet (accessed on 1 October 2021).

Article

Application of Fusion of Various Spontaneous Speech Analytics Methods for Improving Far-Field Neural-Based Diarization

Sergei Astapov [1,*], Aleksei Gusev [1,2], Marina Volkova [1,2], Aleksei Logunov [1,2], Valeriia Zaluskaia [1,2], Vlada Kapranova [1,2], Elena Timofeeva [1], Elena Evseeva [1], Vladimir Kabarov [1] and Yuri Matveev [1,2]

[1] Information Technologies and Programming Faculty, ITMO University, 197101 Saint Petersburg, Russia; gusev-a@speechpro.com (A.G.); volkova@speechpro.com (M.V.); logunov@speechpro.com (A.L.); zaluskaya@speechpro.com (V.Z.); kapranova@speechpro.com (V.K.); eptimofeeva@itmo.ru (E.T.); evseeva@itmo.ru (E.E.); vikabarov@itmo.ru (V.K.); matveev@mail.ifmo.ru (Y.M.)
[2] STC-Innovations Ltd., 194044 Saint-Petersburg, Russia
* Correspondence: sastapov@itmo.ru

Citation: Astapov, S.; Gusev, A.; Volkova, M.; Logunov, A.; Zaluskaia, V.; Kapranova, V.; Timofeeva, E.; Evseeva, E.; Kabarov, V.; Matveev, Y. Application of Fusion of Various Spontaneous Speech Analytics Methods for Improving Far-Field Neural-Based Diarization. *Mathematics* **2021**, *9*, 2998. https://doi.org/10.3390/math9232998

Academic Editors: Grigoreta-Sofia Cojocar and Adriana-Mihaela Guran

Received: 31 July 2021
Accepted: 17 November 2021
Published: 23 November 2021

Publisher's Note: MDPI stays neutral with regard to jurisdictional claims in published maps and institutional affiliations.

Copyright: © 2021 by the authors. Licensee MDPI, Basel, Switzerland. This article is an open access article distributed under the terms and conditions of the Creative Commons Attribution (CC BY) license (https://creativecommons.org/licenses/by/4.0/).

Abstract: Recently developed methods in spontaneous speech analytics require the use of speaker separation based on audio data, referred to as diarization. It is applied to widespread use cases, such as meeting transcription based on recordings from distant microphones and the extraction of the target speaker's voice profiles from noisy audio. However, speech recognition and analysis can be hindered by background and point-source noise, overlapping speech, and reverberation, which all affect diarization quality in conjunction with each other. To compensate for the impact of these factors, there are a variety of supportive speech analytics methods, such as quality assessments in terms of SNR and RT60 reverberation time metrics, overlapping speech detection, instant speaker number estimation, etc. The improvements in speaker verification methods have benefits in the area of speaker separation as well. This paper introduces several approaches aimed towards improving diarization system quality. The presented experimental results demonstrate the possibility of refining initial speaker labels from neural-based VAD data by means of fusion with labels from quality estimation models, overlapping speech detectors, and speaker number estimation models, which contain CNN and LSTM modules. Such fusing approaches allow us to significantly decrease DER values compared to standalone VAD methods. Cases of ideal VAD labeling are utilized to show the positive impact of ResNet-101 neural networks on diarization quality in comparison with basic x-vectors and ECAPA-TDNN architectures trained on 8 kHz data. Moreover, this paper highlights the advantage of spectral clustering over other clustering methods applied to diarization. The overall quality of diarization is improved at all stages of the pipeline, and the combination of various speech analytics methods makes a significant contribution to the improvement of diarization quality.

Keywords: speaker diarization; spontaneous speech processing; voice activity detection; overlapping speech detection; speaker extractor models; speaker number estimation; model fusion; quality estimation; distant speech processing; artificial neural networks

1. Introduction

The widespread availability of tools for sound acquisition, as well as the cost reduction of audio data storage systems, require new methods for automatic processing. Such tasks as the generation of meeting minutes, the processing of telephone conversations, and the automatic transcription of news or entertainment programs, not only require speech recognition [1], but also involve audio annotation by speakers, which is usually referred to as speaker diarization.

Diarization is the process of partitioning an input audio stream into homogeneous segments according to the speaker identity [2]. This means that the goal is to determine who is speaking in each audio segment. Diarization can be used as a preliminary stage in speech recognition systems, in automatic translation or meeting recording transcription.

The choice of a suitable diarization scenario depends on the specific task and corresponding data. So, processed data can be characterized based on the channel specifics (telephone, microphone or microphone array), the presence of noise and the reverberation level, etc. Preliminary information can also influence the choice of diarization methods: whether the exact number of speakers is known or whether individual speech samples of their voices are available. When it comes to meeting minutes, it is helpful to know in advance if the participants can move around the room or tend to interrupt each other. All of these factors can significantly affect the quality of diarization [3].

The diarization scenario may consist of the following steps [4]. The first step usually tends to apply the voice activity detector (VAD) in order to obtain the markup of speech and non-speech segments of an audio recording. This stage can be affected by the quality of the recording or the presence of speaker interruptions, as mentioned above. Nevertheless, the accuracy in the finding of speech boundaries can affect the overall quality of diarization. The next step is the extraction of speaker features (speaker models) for each speech segment. This is necessary for the main purpose of diarization, in order to determine exactly who is speaking in a given segment. At this stage, well-proven representations of the speaker patterns are used: i-vectors obtained via factor analysis [5], x-vectors extracted using a time delay neural network [6] or other types of DNN-embeddings [7]. Having been obtained for each segment, the speaker representations are subject to clustering procedures during the third step of diarization. Pre-selected similarity metrics are used to divide speech segments into clusters and to match the speaker label to each cluster. Thus, in the output of the diarization system, the markup "who speaks when" is obtained. In our paper we consider approaches to improve the quality of diarization at each stage, with particular attention to the detection of speech boundaries.

The accuracy of speech boundary detection can be improved through the choice of an appropriate VAD method. Recently, in addition to standard energy-based voice activity detectors, neural network-based VADs are gaining popularity, which allows one to obtain better resistance to noise conditions [8]. In this paper, we consider the use of the DNN-based VAD described in [9] for a diarization task applied to the AMI Meeting Corpus [10]. To deal with the inherent problems of multi-dialogue recordings such as speaker interruptions and the simultaneous utterances of multiple speakers, we examine options for fusing VAD with other speech analytic systems. At a certain stage we apply instantaneous speaker number estimation, which we further refer to as a speaker counter (SC) model. The model resolves a classification problem, where each class represents the number of simultaneous speakers detected. If the speaker counter is analyzed in terms of two classes, "zero speakers" and "one or more speakers", it can be considered an alternative to VAD, and fusing VAD and SC can increase accuracy in locating speech boundaries. In turn, the performance of SC in conversations with frequent interruptions can be improved by fusing SC with the model which detects overlapping speech segments [11]. We refer to this model as the overlapping speech detector (OSD). The fusion of these two models allows one to tackle the problem of simultaneous speaker detection from separate perspectives.

Information about acoustic conditions in terms of the speech-to-noise ratio (SNR) and reverberation time (RT60) for each audio segment can also be used to distinguish speech and non-speech frames. We apply an automatic quality estimation (QE) system, described in [9], to cluster estimated SNR-RT60 vectors into speech and non-speech clusters and thereby retrieve an approximate voice activity markup. Although this method is not accurate, we investigate its usefulness in fusing it with the base DNN-VAD.

The next stage of the diarization process consists of generating speaker models for each segment located during the previous stages of speech detection. The algorithms at this stage can be implemented through speaker DNN-embedding extraction, similar to the speaker verification task. The current state-of-the-art systems in speaker verification are completely guided by the deep learning paradigm. Previously, the frame-level portion of these extractors was based on TDNN (time delay neural network) blocks that contained only five convolutional layers with temporal context [6]. Such types of embeddings, referred to as x-vectors, are often applied to diarization tasks in state-of-the-art systems [12]. The newer ECAPA-TDNN (emphasized channel attention, propagation and aggregation TDNN) [13] architecture develops the idea of TDNN and contains additional blocks with hierarchical filters for the extraction of different scale features. ECAPA-TDNN and various modifications of the well-known ResNet [7] architecture are compared in our research in terms of EER, minDCF, and DER metrics.

The speaker models obtained for each speech segment must then be clustered. The purpose of clustering is to associate segments from the same speaker to one another. The clustering procedure ideally yields one cluster per each speaker in the recording, with all the segments from a given speaker contained in a single cluster. The common approach used in diarization systems is agglomerative hierarchical clustering, which can be used in the absence of prior knowledge about the number of speakers [14–16]. As an alternative method, spectral clustering [17] can also be used, as well as methods that require the specification of the number of clusters, such as KMeans or DBSCAN [18].

In this study, we aimed to enhance separate parts of the neural-based diarization system and analyze their contributions to the final assessment. In particular, our experiments were based on the following system components: feature extraction, VAD methods, speaker extractor models, speaker verification, and diarization. VAD results coupled with SC and QE estimates were applied to speaker extractor and diarization models' results through fusion algorithms in order to achieve an increase in diarization quality. Thus, separate state-of-the-art and proposed diarization system components, as well as the pipeline in their entirety, are studied and evaluated.

Since the study focuses specifically on cases of noisy overlapping speech acquired by a far-field microphone in real-life conversation conditions, it is important to take into account the same conditions when comparing our solution with those of other studies. In this sense, our proposed neural network-based VAD is compared to the publicly-available SileroVAD [19] on the evaluation subset of the Third DIHARD Speech Diarization Challenge. Our method demonstrated an ability to deal with noisy conditions and showed a 16.9% EER versus the 26.37% EER of SileroVAD. However, it is not always possible to correctly compare the results of different studies. For example, the accuracy of our proposed speaker counter detector, obtained on a realistic AMI dataset, was 65.6%, and, although this percentage was less than those presented in similar works, unlike those works, we did not employ synthesized datasets. As described below, our results in regard to diarization of the AMI evaluation set are comparable to the state-of-the-art papers, and the proposed fusion methods can be further improved.

The paper is structured as follows. Section 2 presents the datasets applied for the training and evaluation of the presented diarization pipeline and its system components and discusses the methods of feature extraction applied in these components. Section 3 describes the separate system pipeline components, namely, the approaches to VAD, SC, OSD, and QE, speaker extraction models, and the fusion methods applied to these system components. Section 4 is devoted to the experimental evaluation of the system components and the diarization pipeline. Finally, Section 5 discusses the results achieved during the study and addresses the prospects of future developments in the considered direction of research.

2. Data Processing

In this section we describe the data used to train and test all systems included in the investigated diarization pipeline. The section begins by describing the process of extracting features from raw audio as the first and primary stage of data processing.

2.1. Feature Extraction

For audio signal processing, we extract several feature types from the raw audio signal that can be fed into machine learning models. In our paper, feature extraction methods vary depending on the diarization system units, the main components of which are speaker embedding extractor, VAD, SC, OSD, and QE models.

All speaker embedding extractors presented in this paper expect log Mel-filterbank (LMFB) energy coefficients extracted from raw input signals using the standard Kaldi recipe [20] at the sampling rates of 8 kHz or 16 kHz:

- consisting of 64 LMFB components extracted from a raw signal with a sampling rate of 8 kHz;
- consisting of 80 LMFB components extracted from a raw signal with a sampling rate of 16 kHz.

Extracted features additionally go through either one of the two different post-processing steps, depending on the type of speaker embedding extractor used afterwards:

- local cepstral mean normalization (CMN-normalization) over a 3-s sliding window;
- global cepstral mean and variance Normalization (CMVN-normalization) over the whole utterance.

The application of the above methods for each type of speaker-embedding extractor is discussed in Section 3.5.

To obtain speech segments from the audio signal, we apply a neural network-based voice activity detector (VAD) system developed by us. The model receives mel-frequency cepstral coefficients (MFCCs) extracted from the raw signal with a sampling rate of 8 kHz.

Modified versions of the voice activity detection system integrate OSD, SC or QE models with DNN-VAD. As the features, the OSD and SC neural network models use the short-term Fourier transform (STFT) coefficients extracted from the preprocessed input audio with a sampling rate of 16 kHz. The QE models use LMFB feature coefficients with CMN-normalization. The parameters of the extracted features are presented in Table 1.

Table 1. Types of applied audio features.

Model	Feature Type	Number of Coefficients	Frame Length, ms	Overlap, ms	Sampling Rate, kHz
Speaker Embedding Extractors	LMFB	64/80	25	15	8/16
VAD	MFCC	23	30	10	8
SC	STFT	201	25	10	16
OSD	STFT	81	25	10	16
QE	LMFB	64	25	15	8

2.2. Datasets

In this paper, two different sets are used for the test protocol. The target dataset for measuring the quality of diarization is the AMI corpus. This dataset is also used to train SC and OSD models, evaluate individual parts of the system (SC, OSD, clustering) and fusion. Additionally, during the study, Voxceleb1 and NIST SRE 2019 evaluation datasets are used to measure the quality of speaker verification models. In the current work, we decided to investigate the degradation of speaker extraction systems trained on the utterances sampled at 8 kHz, compared to the systems trained on files sampled at 16 kHz. For this purpose, two different datasets were created. For all datasets, the division into train/dev/test sets suggested by their original authors was used. The application of each of these datasets is discussed in detail below.

The main dataset for assessing the quality of diarization is the **AMI corpus** [10]. The dataset consists of over 100 h of meeting recordings. In general, the total number of participants in a single meeting is four (approximately 80%), and rarely three or five. The meetings were recorded in specially equipped rooms of various configurations and locations using microphone arrays, lapel microphones and headphones. In addition, each meeting attendee was provided with graphics such as videos, slides, whiteboards, and notebooks. All recordings were synchronized. The dataset contains both the recordings of real meetings and meetings with predefined scenarios and roles. For about 70% of the recording duration only one speaker is active, whereas for about 20% of the duration speech is absent, and only 10% corresponds to the simultaneous speech of several people.

Several recent studies on the diarization problem include experiments based on the AMI corpus to measure quality. The proposed methods can differ in speaker extractor models, clustering methods, whether they include or exclude overlapping speech in the scoring, as well as whether they use references or predicted speech/non-speech labels. It is also essential to use the same evaluation protocols for the AMI database for a fair comparison, in particular, to select data from the same set of microphones: Headset-Mix, Lapel-Mix or Distant-Mic. The work [21] closest to our solution compares the x-vector TDNN and ECAPA-TDNN architectures for speaker model extraction and ignores overlapping speech segments during scoring. In this work, the best results are obtained using the ECAPA-TDNN architecture with a spectral clustering back-end, which achieved 3.01% DER for the case of the estimated number of speakers and 2.65% DER for the case of a known number of speakers on the AMI Headset-Mix evaluation set. Another work [22] compares the well-known agglomerative hierarchical clustering (AHC) method and a proposed modification of the variational Bayes diarization back-end (VBx) method, which clusters x-vectors using a Bayesian hidden Markov model (BHMM). The AHC in their experiments showed 3.96% DER, whereas VBx with a single Gaussian model per speaker showed 2.10% DER for the AMI Headset-Mix evaluation set. Since the analysis in both of the abovementioned papers focuses on oracle VAD, they can be comparable with our results, presented in Section 4.3.

In our research, we applied full-corpus-ASR [10] partitioning of meetings and used the evaluation part of the lapel-mix AMI corpus for diarization experiments and system fusion. For the training and evaluation of the speaker counter and overlapping speech detector models the training and evaluation parts of the Array1-01 AMI corpus were used, respectively. Additionally, the following datasets were used during the intermediate steps of our proposed approach.

The **Voxceleb1** dataset [23,24] is composed of audio files extracted from YouTube videos and contains 4874 utterances recorded at 16 kHz. The speakers span a wide range of different ethnicities, accents, professions and ages. Segments include interviews from red carpets, outdoor stadiums and indoor studios, speeches given to large audiences, excerpts from professionally-shot multimedia, and even crude videos shot on hand-held devices. Crucially, all are degraded with real-world noise, consisting of background chatter, laughter, overlapping speech, and room acoustics [25]. The quality of the recording equipment and channel noise quantity also vary quite noticeably.

The **NIST SRE 2019 evaluation** dataset [26] is composed of PSTN and VoIP data collected outside of North America, spoken in Tunisian Arabic, and contains 1364 enrollment and 13,587 test utterances recorded at 8 kHz. Speakers were encouraged to use different telephone instruments (e.g., cell phones, landlines) in a variety of settings (e.g., a noisy cafe, a quiet office) for their initiated calls [26]. Enrollment segments approximately contain 60 s of speech to build the model of the target speaker. The speech duration of the test segments is further uniformly sampled with lengths varying from approximately 10 s to 60 s.

The **16 kHz training set**. For this set we concatenated VoxCeleb1 and VoxCeleb2 (SLR47) [25] corpora. We used videos from VoxCeleb1 and VoxCeleb2, and concatenated all the corresponding audio files into one chunk. Augmented data were generated using the standard Kaldi augmentation recipe (reverberation, babble, music and noise) using the freely available MUSAN and simulated room impulse response (RIR). In total, the training dataset contains 833,840 recordings from 7205 speakers.

The **8 kHz training set**. For this set we used a wide variety of datasets, containing telephone and microphone data from private datasets and from those available online. The dataset includes Switchboard2 Phases 1, 2 and 3, Switchboard Cellular, Mixer 6 Speech, data from NIST SREs from 2004 through 2010 and 2018, concatenated VoxCeleb 1 and 2 data, extended versions of the Russian speech subcorpus named RusTelecom v2 and the RusIVR corpus. RusTelecom is a private Russian speech corpus of telephone speech, collected by call centers in Russia. RusIVR is a private Russian speech corpus containing speech, collected in different scenarios, such as noisy microphones, telephone calls, recordings from distant recorders, etc. All files are sampled at 8 kHz. In order to increase the amount and diversity of the training data, augmentation using the standard Kaldi augmentation recipe (reverberation, babble, music and noise) was applied using the freely available MUSAN and simulated room impulse response (RIR) datasets. In total, this training dataset contains 1,679,541 recordings from 33,466 speakers.

3. Methods

In this section a detailed description of the main systems included in our diarization pipeline is provided. The first four subsections (VAD, SC, OSD and QE) describe the models used in various types of speech boundary detection and the fusion of these models. Section 3.5 is devoted to speaker extraction models and the methods used for their training.

3.1. Voice Activity Detection

This work proposes a fusion method of three different models, each of which shows different quality on the same evaluation subset of the AMI corpus. The first one is the voice activity detector (VAD), which is trained purposefully for the task of speech boundary detection on the AMI corpus. In our case, we use the method proposed in [7], which adapts the idea of using the U-Net architecture for segmentation from the spatial to the time domain. This architecture was originally introduced in [27], as a fairly precise method for object localization in microscopic images. U-Net is a convolutional architecture, that involves the idea of the deconvolution of small and deep image representation with many small layers into an image of the original size by applying the upsampling operation. In this study, we apply a reduced version of the original U-Net architecture, which is presented in Figure 1. Since the task of detecting speech activity is a task of segmentation in the time domain, we apply the combination of Dice and cross-entropy losses as the main loss-function in the VAD model training process [28].

The training process pipeline of the model for the AMI task consists of two stages:
1. Fitting on the main training set;
2. Adaptation on the AMI corpus.

During the first stage of training we use the concatenation of the NIST 2002/2008 speech recognition datasets and the RusTelecom corpus, described in Section 2.2. This data setup leads to a confident VAD quality of about a 10% equal error rate (EER) on different configurations of the model. Adaptation on the AMI corpus is described via the same training process as during the first stage, but using a smaller learning rate for the fitting of new data without the loss of already learned knowledge. In general, and also in our specific case, the adaptation process should last for a small amount of training iterations to prevent the overfitting on the new adaptation dataset and the reduction of the previous ability to detect speech in common.

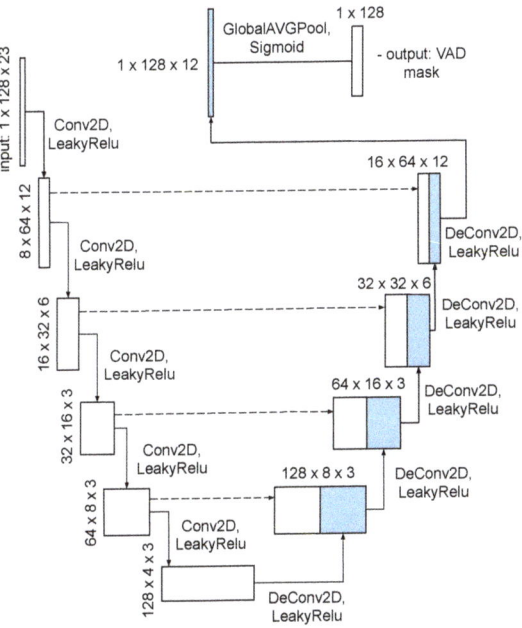

Figure 1. The applied U-Net VAD model architecture.

To compare VAD performance under the same conditions, we computed EER scores on the evaluation subset of the Third DIHARD Speech Diarization Challenge using our VAD and a similar method called SileroVAD [19]. The results were 16.9% EER using our method and 26.37% using SileroVAD.

3.2. Speaker Counter Model

Another model, the results of which can be interpreted as a time-domain speech detection markup, involves a detector of the number of simultaneous speakers, known as a speaker counter (SC). Theoretically, the estimation of the number of concurrent speakers is closely related to the problem of speaker identification, which is considered one of the tasks involved in speaker diarization [2,29–31].

We have compared the results of our model with the state-of-the-art systems in speaker number estimation presented in [11,32], which are based on convolutional neural networks. We have taken into consideration the models from the cited papers and the model presented in the current work, trained on 1-s recording segments to count the number of speakers. In [32], the authors present a model trained and evaluated on a synthetic dataset, which performs the speaker counting task with an average F1-score of 92.15% for classes with 0–3 speakers. In [11], the authors present a model trained and evaluated on mixtures of speaker recordings of the LibriSpeech dataset; this model achieves 77.3% accuracy for classes with 1–4 speakers.

Previously referenced works in the field mainly consider the synthetic mixtures of different speaker recordings and thus obtain permissible results for active speaker number estimations. We, however, focus specifically on real-life recordings of natural conversations that are contained in the AMI corpus, which reduces estimation quality. Our study has shown that no similar works in the field contain results based specifically on the AMI corpus; thus, exact comparison with our results is not possible. To train the model we applied data augmentation to the AMI corpus, and tested the model on the evaluation set of this corpus. The model achieved an F1-score of 65.6% in real-speech conditions. The results are further discussed in Section 4.4.

The SC model solves the task of estimating the number of concurrent speakers, which is formulated as a classification task with 5 classes, from 0 to 4+ concurrent speakers. The model is based on the architecture described in [33]. The solution is based on the application of several deep neural network (DNN) models, a diagram of which is presented in Figure 2. These models consist of several convolutional layers, max-pooling layers, a Bi-LSTM layer with 40 hidden units for processing input features, and a fully connected layer, which implements the classification of the number of active speakers. The softmax function is used for the output layer; thus, the prediction is specified by the highest probability of the output distribution [34].

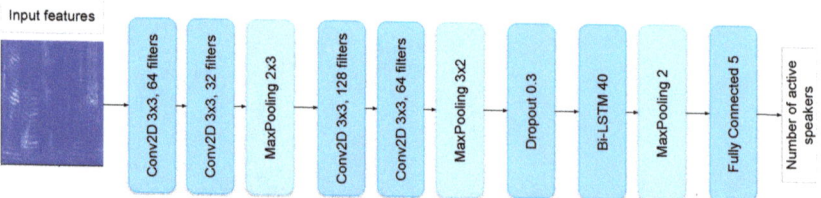

Figure 2. Speaker counter and overlapping speech detector model architecture.

To create the markup that is applicable to the SC model, we employed the manual annotations provided with the AMI corpus. These annotations specify the speech segments for each speaker. To obtain the markup with the number of active speakers, the beginning and end time stamps of all speech segments are combined into one set and sorted. This sorted set is then divided into subsegments, and for each subsegment the number of concurrent speakers is calculated using the original annotation [34]. Thus, we extract the subsegments where the number of simultaneous speakers remains unchanged, from 0 (silence) to 4+ overlapping speakers in a subsegment.

3.3. Overlapping Speech Detection

The model for the task of overlapping speech detection is similar to the SC model in its architecture (see Figure 2), but employs a different type of speech markup. The overlapping speech detector (OSD) is applied for the detection of overlapping speech segments in a conversation involving multiple speakers [11,35–40]. The OSD performs a binary classification task of detecting overlapping and non-overlapping speech. The markup for this task is produced using the speaker number markup of the SC task:

$$l_{OSD} = \begin{cases} 0, & l_{SC} \in \{0,1\}, \\ 1, & l_{SC} \in \{2,3,4+\}, \end{cases} \quad (1)$$

where l_{OSD} is an OSD segment class label and l_{SC} is an SC segment class label. The model presented in Figure 2 is trained with the OSD labels to perform the detection of overlaps. The model accepts the features extracted from short audio segments as input, whereas the output corresponds to the probabilities of the classes overlapping/non-overlapping per each speech segment.

3.4. QE-Vectors System

As an alternative method of separating speech and non-speech segments, we used the information about the acoustic characteristics of the signal from the microphone array of the AMI corpus. During evaluation, for each 0.5 s signal frame we estimated the SNR and RT60 parameters using the signal quality estimation (QE) model described in [9]. SNR and RT60 parameters were predicted for all 8 channels of the microphone array, so for each 0.5 s frame we obtained a vector of 16 values. The resulting vectors were clustered using the

K-means method into two clusters. To decide which of the clusters represented the speech segments and which represented non-speech, the centers of each cluster were compared with the mean speech vector, which is calculated in advance for the speech segments of the development part of the AMI database. This way it was possible to retrieve the markup of speech and non-speech segments for further fusing with a more accurate method, such as base DNN-VAD.

3.5. Speaker Extractor Models

This section contains the descriptions of the foremost architectural details, training techniques, and distinctive features of the implemented speaker verification models. The most common approaches for speaker verification today are based on blocks with one-dimensional (TDNN) or two-dimensional (ResNet) convolutions, which influenced our choice regarding the speaker extractors' architecture. Thus, in this paper, we explore the ECAPA-TDNN architecture, which develops the idea of an x-vector TDNN, and ResNet-based models, trained on 8 kHz and 16 kHz datasets. We used ResNet34 as a baseline model and attempted to increase the quality of the speaker model by increasing the depth and width of the layers and by investigating the model architectural configuration.

ECAPA-TDNN. Emphasized channel attention, propagation and aggregation in TDNN (ECAPA-TDNN), proposed in [13], is a modification of the standard time delay neural network (TDNN) architecture, containing squeeze-excitation (SE) blocks and Res2Net modules in the frame level with hierarchical filters for the extraction of features of different scales. To process signals of arbitrary duration, the architecture uses attentive statistic pooling instead of the usual statistic pooling. For this work, we chose the official pre-trained model from https://github.com/speechbrain/speechbrain (accessed on 2 March 2021), which was trained on the original VoxCeleb1 and VoxCeleb2 datasets with SpecAugment [41], speed perturbation, noise, and reverberation augmentation. In detail, the model configuration and training are described in [42].

ResNet34. This extractor is based on the ResNet34 model with some modifications, as well as the Maxout activation function on the embedding layer, set to one stride in the first BasicBlock and changed to a simple Conv2D stem block. This model was trained on the 16 kHz dataset with local CMN-normalization and global CMVN-normalization, sequentially. During the training process, the extractor handles short audio segments with the length fixed at 2 s using AM-Softmax loss. During the training stage, the parameters m and s were set equal to 0.2 and 30, respectively. The learning rate was set to 0.001 for the first two epochs, and then it was decreased by a factor of 10 for each consecutive epoch. Detailed information about the model configuration is presented in [7].

ResNet72. For this extractor, the ResNet architecture from [43] was adapted to the task of speaker verification. The model was trained on the 16-kHz dataset with local CMN-normalization. As a loss function, we used adaptive curriculum learning [44]. Parameters m and s were respectively equal to 0.35 and 32 during the whole training stage. The optimization process was based on the SGD optimizer with the momentum of 0.9 and weight decay of 10^{-5} for the head and 10^{-6} for the rest of the layers. Moreover, for learning rate control the OneCycleLR scheduler was used with the maximum learning rate fixed to 2. To increase the training speed, we also used AMP (Automatic Mixed Precision) with half-precision, which raised the batch size per GPU. The model was trained during 15 epochs on randomly sampled 4-s crops of each utterance from the training dataset. The comprehensive description of the model configuration is presented in [45].

ResNet101_8k. The ResNet101 model modification included the Maxout activation function on the embedding layer, set to one stride in the first BotleneckBlock and changed to a simple Conv2D stem block, which provided the basis for this extractor. The training set for this model was the 8 kHz dataset with local CMN-normalization performed in several stages. In the first stage, the model was trained for 20 epochs on randomly sampled 6-s crops of each training dataset utterance. AAM-Softmax losses with parameters m and s were respectively equal to 0.35 and 32. In the second stage, crop duration was increased to

12 s and the parameters of AAM-Softmax loss were set to 0.45 and 32, respectively. Then, the model went through 10 epochs of training with these settings.

ResNet101_48_8k. The ResNet101_8k model was set to be used directly as a basis for the extractor at hand, with the extension number of convolution layers in the model architecture ranging from 32 N to 48 N, where $N \in \{1, 2, 4, 8\}$ for different model blocks. Higher quality was achieved through an increase in the size of convolutional layers; however, the increase in the number of model parameters resulted in an increase of training time compared to the ResNet101_8k model.

3.6. Model Fusion

In order to improve the accuracy of speech boundary detection, we conduct a series of experiments by fusing the speech markups obtained using DNN-VAD with the markups obtained using other systems discussed above: SC, OD and QE. The following options for fusion were considered:

- VAD + QE, on the assumption that information about acoustic conditions (RT60, SNR) can correlate with the speech component properties and thereby improve VAD markup;
- VAD + SC, on the assumption that the above-described model for counting the number of speakers can be considered as an alternative approach to the detection of speech activity;
- VAD + QE + SC + OD, on the assumption that the SC + OD fusion is shown to produce good quality estimates (see Section 4.4), and the combination of different approaches in the conditions of a multispeaker conversation can improve the accuracy of voice annotation compared to baseline DNN-VAD.

According to the diarization pipeline, the obtained speech markup was then used for speaker model extraction; so, the accuracy of speech boundary detection affected the overall diarization quality. We describe the details of the applied voice annotation system fusion and analyze the influence of different fusion approaches on speaker diarization quality in terms of DER in Section 4.5.

4. Experiments

Our main experiment was aimed at the fusion of different speech analytics techniques to improve the overall quality of diarization. A description is available in Section 4.5, which includes the comparison of different combinations of speech markup methods in terms of DER: VAD, QE, SC and OSD, outlined above. In the final experiment we intended to apply the speaker extractor model chosen in Sections 4.1 and 4.2, the selected clustering method (Section 4.3), and the SC + OSD fusion method (Section 4.4) in combination.

Thus, Sections 4.1–4.4 describe preliminary experiments aimed at independently improving each of the stages of fusion. Section 4.1 contains the comparison of different speaker extractor models in the speaker verification task. We tested the generalizability of different neural network architectures trained on the 16 kHz training set and the combined 8 kHz training set described in Section 2.2 to cope with unknown evaluation data. Section 4.2 compares the same speaker extractor models for an AMI-based diarization problem using ideal VAD markup. As a result of these experiments, we defined the ResNet101_8k model as the most optimal, so it was chosen for the final experiments. Section 4.3 is devoted to the next stage of the diarization pipeline—the comparison of clustering methods for extracted voice models of speakers. The experiments were carried out on the ResNet101_8k and ECAPA_TDNN models. Finally, Section 4.4 focuses on SC and OSD fusion to refine speech boundaries under interruption conditions.

4.1. Speaker Verification

We investigated the quality of different speaker extractors for the speaker verification task on the Voxceleb1 test and NIST SRE 2019 eval sets. Since the speaker verification system receives two sets of speech recordings during the evaluation process, originating either from the same speaker or from two different speakers, there arise two types of errors: the false acceptance rate (FA) and the false rejection rate (FR), which are dependent on the decision threshold. Then, the equal error rate (EER) is the point at which both rates are equal to each other. The lower the EER value, the higher the overall accuracy of the biometric system [46].

In order to take into account the different FR and FA error costs, there exists the detection cost function (DCF or C_{det}) measure, which is described by the following equation:

$$C_{det}(\theta) = C_{FR} \cdot P_{tar} \cdot P_{FR}(\theta) + C_{FA} \cdot (1 - P_{tar}) \cdot P_{FA}(\theta), \qquad (2)$$

where C_{FR} and C_{FA} are the estimated cost parameters for false accept and false reject errors, P_{tar} is a prior probability of targets in the considered application; error-rates P_{FR} and P_{FA} are determined by counting errors during evaluation, and θ is the decision threshold.

The optimal value of DCF obtained via the adjustment of the detection threshold is minDCF. In our experiments we used the EER measure and minDCF with the *a priori* probability of the specified target speaker (P_{tar}) set to 0.01, $C_{FR} = 1$, $C_{FA} = 1$.

The main idea of this experiment consisted of comparing the models trained on different types of data (8 kHz or 16 kHz) and tested on the test subsets of the same data (8 kHz or 16 kHz as well). For this task, we upsampled the 8 kHz audio files to 16 kHz or downsampled the 16 kHz audio files to 8 kHz. For ECAPA-TDNN, we used the official implementation of the model and data processing pipeline from [42]. Note that VAD was not used for feature processing for ECAPA-TDNN, according to the official implementation. No normalization or adaptation techniques were used for speaker embedding comparisons. Simple cosine similarity between speaker embedding vectors was used as a score for speaker model comparisons. The results of the comparison between the speaker verification models are presented in Table 2. To determine the confidence intervals for the EER estimates, we used the fast equal error rate confidence interval (FEERCI) algorithm to calculate non-parametric, bootstrapped EER confidence intervals [47]. The main idea of this approach is to use random subsets from genuine and imposter score lists for estimation. We used 95% confidence intervals on 50 bootstrap iterations for this task. We estimated the confidence intervals for the fixed architecture of the model and its weights due to the significant computational complexity of the task of retraining the speaker recognition model. Based on the results, the following conclusions can be drawn:

- The system trained for specific data types works better on the test set of the corresponding type. Data type mismatches led to significant quality degradations for all tested systems;
- The quality degradation of models trained on a combined dataset (ResNet101_8k, ResNet101_48_8k) was less than the degradation of the models trained only on VoxCeleb datasets. This can be explained by the growth of the generalizing ability of the network with the increase in samples of the training dataset;
- The ResNet-based models trained on VoxCeleb datasets showed better quality on out-of-domain tasks, compared to ECAPA-TDNN, which can be observed by comparing the results of speaker verification. However, ECAPA-TDNN shows a better result on the in-domain test dataset.

Table 2. Results of speaker verification systems for the in-domain Voxceleb1 test and the out-of-domain NIST SRE 2019 eval sets in terms of the EER ± 95% confidence interval/minDCF (%). Lower error values are better.

Model	Train Set	EER ± ci = 0.95/MinDCF0.01 (%)	
		VoxCeleb1 Test	NIST SRE 2019 Eval Set
ECAPA-TDNN	VoxCeleb1, VoxCeleb2 [13]	0.71 ± 0.09/0.092	14.52 ± 0.11/0.785
ResNet34	16 kHz	1.18 ± 0.10/0.126	14.37 ± 0.12/0.748
ResNet72	16 kHz	1.11 ± 0.07/0.093	12.75 ± 0.11/0.723
ResNet101_8k	8 kHz Combined	1.54 ± 0.15/0.156	2.97 ± 0.07/0.276
ResNet101_48_8k	8 kHz Combined	1.43 ± 0.14/0.135	2.85 ± 0.07/0.280

4.2. Diarization

The quality of various speaker verification systems was compared during the task of diarization. The embeddings of each continuous speech segment were extracted with the chosen sliding window and shift duration values. We investigated the influence of these parameters based on the results of diarization in terms of the diarization error rate (DER), which consists of three types of error: speaker error, missed speech, and false alarm. DER is denoted as

$$DER = E_{spkr} + E_{miss} + E_{fa}, \quad (3)$$

where E_{spkr} is the speaker error, the percentage of the scored time when a speaker ID is assigned to the wrong speaker. This type of error does not account for overlapping speakers or any error situated within non-speech frames. E_{miss} is missed speech—the percentage of scored time when a hypothesized non-speech segment corresponds to a reference speaker segment. E_{fa} is false-alarm speech, the percentage of scored time when a hypothesized speaker is labeled as non-speech in the reference annotation [48].

We used the DER metric configured according to NIST: a forgiveness collar of 0.25 s was used and the speaker overlap regions were ignored during scoring. We used the ideal VAD markup computed from the ground truth information from the AMI dataset. The results of the comparison of the speaker verification models in terms of DER are presented in Table 3. The development set was used for tuning the spectral clustering parameters. The tuned parameters were used to perform the diarization task on the evaluation set.

Table 3. Results of investigated systems for AMI Development and Evaluation sets (dev/eval) in terms of DER (%) depending on the set sliding window and shift duration values. Lower error values are better. Upper and lower 0.95 confidence intervals are indicated with a dash.

Model	DER (%) on dev with ci = 0.95			
	Win = 1.0 s Shift = 0.5 s	Win = 1.5 s Shift = 0.75 s	Win = 2.0 s Shift = 1.0 s	Win = 3.0 s Shift = 1.5 s
ECAPA-TDNN	3.90–5.29–6.14	1.95–2.50–2.84	1.65–2.17–2.46	1.92–2.37–2.65
ResNet34	3.51–5.12–6.09	1.48–2.35–2.76	1.54–**1.78**–2.03	1.74–2.65–3.05
ResNet72	2.48–**3.39**–3.96	1.35–2.10–2.44	1.38–1.91–2.21	1.72–**2.12**–2.38
ResNet101_8k	2.44–**3.38**–3.89	1.46–2.12–2.43	1.30–1.90–2.15	1.76–2.28–2.56
ResNet101_48_8k	2.56–3.99–4.72	1.40–**1.88**–2.16	1.42–**1.80**–2.04	1.82–**2.15**–2.43

Model	DER (%) on eval with ci = 0.95			
	win = 1.0 s shift = 0.5 s	win = 1.5 s shift = 0.75 s	win = 2.0 s shift = 1.0 s	win = 3.0 s shift = 1.5 s
ECAPA-TDNN	3.44–3.99–4.52	2.45–2.81–3.15	2.15–2.42–2.67	2.52–2.81–3.10
ResNet34	2.93–3.73–4.39	1.93–2.46–2.91	2.00–3.20–3.85	2.05–2.33–2.78
ResNet72	2.92–**3.59**–4.32	1.77–1.94–2.30	1.41–**1.58**–1.85	2.02–2.56–3.05
ResNet101_8k	3.52–3.90–4.61	1.85–2.02–2.36	1.62–1.81–2.10	1.99–**2.17**–2.53
ResNet101_48_8k	2.97–3.66–4.19	1.63–**1.75**–1.95	1.47–1.64–1.82	2.00–2.23–2.52

To determine the confidence intervals for the DER estimates we used the following algorithm:

1. Input: $(\{w_i, \ldots, w_n\})$; w—diarization results for each file in subset, n—number of unique files in subset;
2. Choose random subsets of size $n-2$ from the entire subset and compute the DER metric for the files in each of $n-2$ size subsets. We use $n-2$ files to maximize the size of subset, while assuring a sufficient variability of subsets;
3. Repeat the 2nd step 50 times, sort all computed DER results, compute mean and 95% confidence interval thresholds.

Based on the presented results, the following conclusions were drawn:

- The quality of the ResNet-based model outperformed ECAPA-TDNN in the diarization task for the AMI dataset;
- The best quality on the evaluation set of AMI was achieved using the sliding window and shift duration values set to 2.0 s and 1.0 s. respectively;
- High-quality diarization was achieved using the model trained on the 8 kHz dataset, indicating that the data variability of the training dataset was no less important for achieving better diarization results than the type of train data. The quality of the model trained on the 8 kHz dataset matched or even exceeded the quality of models trained on test-like data;
- The sizes of the development and evaluation parts of the AMI dataset were probably insufficient for quality assessments with high confidence intervals. For example, the quality of all systems on the evaluation set for sliding window and a shift duration values of 1.5 s and 0.75 s, accordingly, were comparable in the confidence interval. Furthermore, the quality of ResNet72, ResNet101_8k and ResNet101_48_8k models for the window and shift parameters larger than the abovementioned values were comparable.

Here and in other tables of this paper, some metrics contains range values which correspond to boundaries of confidence intervals. These values are separated with a minus sign ("−").

4.3. Clustering

This section compares the clustering algorithm proposed in [17,21] with other clustering methods from Kaldi and sklearn (https://scikit-learn.org/stable/ accessed on 18 May 2021) libraries. For this set of experiments, we used the ResNet101_8k speaker verification system with the best configuration for this model, as discussed in Section 4.2. First, we applied clustering methods with a reference number of clusters, where the number of speakers was computed based on original AMI annotation.

1. Spectral Clustering with cosine affinity and unnormalized modification;
2. Spectral Clustering with cosine affinity and without unnormalized modification from the sklearn package;
3. The k-means clustering algorithm from the sklearn package;
4. Gaussian mixture model with tied covariance matrices, trained on speaker-embedding vectors. The number of used gaussians was chosen according to the reference number of speakers;
5. The density-based spatial clustering of applications with noise (DBSCAN) algorithm was tested but a stable configuration yielding sufficient quality was not achieved; thus, the results of the method at hand are not presented in Table 4.

Second, several clustering algorithms with automatic estimation of the number of clusters were compared.

1. Spectral clustering with cosine affinity and unnormalized modification. The number of speakers was estimated using the maximum eigengap between eigenvalues, computed on the Laplacian matrix, which was calculated based on the pruned cosine similarity matrix between speaker-embedding vectors [21].

2. Agglomerative clustering with cosine affinity and average linkage. The sklearn and Kaldi implementations of this clusterization were used, with insignificant differences between the results in terms of DER.

All the results achieved for the clustering methods are presented in Table 4. Based on these results, better performance was achieved using an unnormalized modification of the spectral clustering algorithm for both cases of the specified reference number of speaker clusters and for the estimated number of clusters.

For the visualization of the markups produced by different speaker models, we used t-distributed stochastic neighbor embedding (TSNE). TSNE results for the ResNet101_8k and the ECAPA-TDNN models are presented in Figures 3 and 4, respectively. The original speaker embeddings and their decomposed representations after spectral decomposition performed internally by SC are visualized. We used the ideal clustering markup with removed speech segments of overlapping speech, estimated after K-means testing for the original speaker embeddings after SC clustering. Note that for both models the results of DER for this file consisted of approximately 1–2% of samples.

Table 4. Results of comparing the clustering algorithms for the AMI Development and Evaluation sets (dev/eval) in terms of DER (%) for the reference and the estimated number of speakers in each single recording. The methods are numbered according to their presentation in Section 4.3. Lower error values are better.

Clustering	Reference Speakers	Dev	Eval
1. Spectral Clustering unnorm	True	1.91	1.81
2. Spectral Clustering	True	2.18	3.05
3. Kmeans	True	2.31	4.38
4. GaussianMixture	True	2.39	4.79
1. Spectral Clustering unnorm	False	2.13	2.78
2. Agglomerative Clustering	False	3.64	3.83

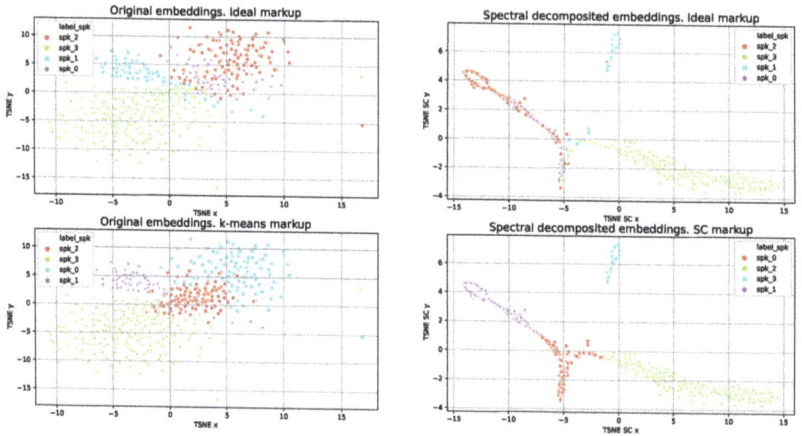

Figure 3. Results of TSNE visualization computed by means of the ResNet101_8k model and EN2002b lapel-mix file from the AMI evaluation dataset. All overlapping speech segments were removed to reduce uncertainty. In the **left** parts of the images, the original speaker embeddings are used; in the **right** parts the speaker embeddings after spectral decomposition are used.

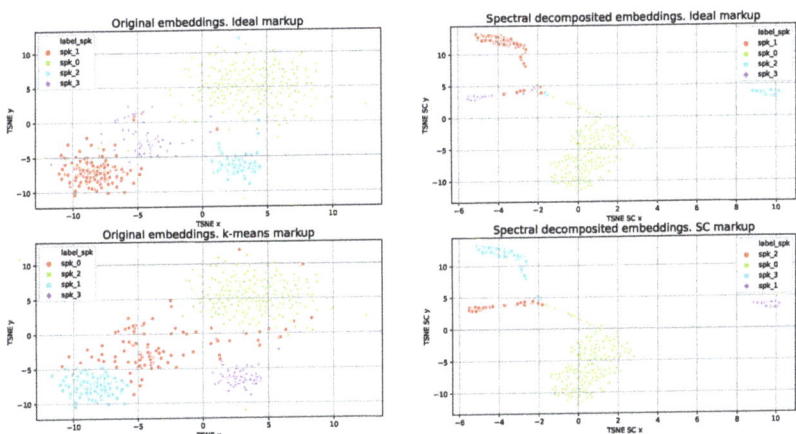

Figure 4. Results of TSNE visualization computed by means of the ECAPA-TDNN model and EN2002b lapel-mix file from the AMI evaluation dataset. All overlapping speech segments were removed to reduce uncertainty. In the **left** parts of the images, the original speaker embeddings are used; in the **right** parts the speaker embeddings after spectral decomposition are used.

4.4. Fusion of SC and OSD Models

The fusion of the SC and OSD models was proposed to improve the quality of diarization during simultaneous speech. Combining these models, the SC performance can potentially be improved at the points of overlapping speech (Figure 5). In a case in which the SC model does not perform well for the speaker number classes of 2, 3 and 4+ simultaneous speakers, but the binary problem of overlap/non-overlap detection yields sufficient estimation quality, the performance of SC may be tuned according to OSD estimates [34]. If SC underestimates the number of speakers, marking most samples as one speaker, then, by fusing SC with OSD the samples with one speaker can be excluded, eliminating the classification error for samples with more than two reference speakers estimated as one speaker. On the other hand, SC may overestimate the number of speakers. In this case, by fusing the models and excluding samples containing 2–4+ speakers, the estimation quality for zero and one speakers can be improved [34].

The principle of the proposed fusion method lies in adjusting the SC class probabilities based on the decision of OSD [34]. If the decisions of the two models are logically in concurrence with one another, the SC probabilities are increased by applying a weight coefficient α. On the other hand, if the decisions of the two models are not in concurrence, the SC probabilities are reduced. Specifically, if OSD defines a speech segment as non-overlapping, then this estimate is used to increase SC probabilities for labels 0 and 1 (no active speaker, one active speaker), and reduce the probabilities for labels 2, 3, and 4+ (2, 3, 4+ active speakers). For segments with estimated overlapping speech, the SC probabilities for labels 2, 3, 4+ are increased, and reduced for labels 0 and 1. The fusion process is generalized by means of the following equation [34]:

$$P_{fusion} = \begin{cases} P_{SC}(l_{SC}) + \alpha \cdot (1 - P_{SC}(l_{SC})), & \hat{l}_{OSD} = 0, \hat{l}_{SC} \in \{0,1\}, \\ P_{SC}(l_{SC}) - \alpha \cdot P_{SC}(l_{SC}), & \hat{l}_{OSD} = 1, \hat{l}_{SC} \in \{0,1\}, \\ P_{SC}(l_{SC}) - \alpha \cdot P_{SC}(l_{SC}), & \hat{l}_{OSD} = 0, \hat{l}_{SC} \in \{2,3,4+\}, \\ P_{SC}(l_{SC}) + \alpha \cdot (1 - P_{SC}(l_{SC})), & \hat{l}_{OSD} = 1, \hat{l}_{SC} \in \{2,3,4+\}, \end{cases} \quad (4)$$

where $P_{SC}(l_{SC})$ are the probabilities of l_{SC} labels for a given segment, $\hat{l}_{OSD}$ and $\hat{l}_{SC}$ are the OSD and SC estimates, respectively, and α is the weight coefficient. The value for α was chosen to be equal to 0.5 during our fusion experiments.

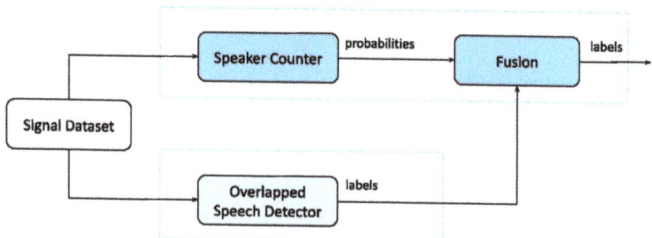

Figure 5. Diagram of speaker counter and overlapping speech detector fusion.

Table 5 presents the results for each class, as well as their weighted average in the metrics of accuracy, completeness, and F1-score. We used the following equation to calculate the F1-score metric, which combines the values of precision and recall; the equation was applied as in the Python module scikit-learn (https://scikit-learn.org/stable/modules/generated/sklearn.metrics.f1_score.html accessed on 14 June 2021):

$$\text{F1-score} = 2 \cdot \frac{precision \cdot recall}{precision + recall}. \quad (5)$$

Table 5 also presents the confidence intervals for the F1-score. To calculate the confidence interval, we take the model results calculated on all files, randomly select 80% of labels, calculate the F1-score for them, and repeat this procedure 50 times. Then we compute 95% confidence intervals by applying the Python module scipy.stats (https://docs.scipy.org/doc/scipy/reference/stats.html accessed on 21 June 2021), taking into account the fact that the obtained F1-score values belong to a normal distribution.

The SC model fused with OSD yields better speaker number estimation quality compared to the non-fused SC model. Experimental results show an increase in recognition quality for all classes, in particular for labels 1 and 2. The recall metric shows no increase, probably due to the fact that SC is initially efficient in classifying segments with numbers of speakers larger than two. The recall increases only when the number of speakers is equal to 1. This means that SC often tends to overestimate the number of speakers. A noticeable increase occurs in the precision metric. This indicates that SC has become more efficient in cases where the number of speakers is equal to 1, whereas previously it tended to overestimate the number of speakers in this case.

Table 5. Speaker counter and overlapping speech detector fusion results.

Method	Class	Precision	Recall	F1-Score ± ci = 0.95
SC	0	0.621	0.755	0.682 ± 0.003
	1	0.736	0.593	0.657 ± 0.001
	2	0.274	0.390	0.322 ± 0.002
	3	0.196	0.242	0.216 ± 0.003
	4+	0.136	0.097	0.113 ± 0.008
	weighted	0.593	0.546	0.562 ± 0.001
SC + OSD	0	0.639	0.744	0.688 ± 0.002
	1	0.781	0.772	0.777 ± 0.001
	2	0.400	0.412	0.406 ± 0.002
	3	0.295	0.222	0.253 ± 0.004
	4+	0.163	0.093	0.119 ± 0.008
	weighted	0.654	0.659	0.656 ± 0.001

4.5. Fusion of Voice Annotation Methods

In the following experiments, we investigated the system VAD markup and reference number of speakers used for diarization. We focused on ways to improve the speech markup by fusing the results of various algorithms presented earlier in this study. For this purpose, a better configuration of VAD was obtained in terms of DER for the AMI evaluation dataset. We used results estimated by applying only the VAD markup as a baseline for other fusion results. We used the ResNet101_8k speaker verification system with the best configuration for the construction of diarization vectors using the system VAD markup. For each segment of the source utterance, the fusion algorithm computed a simple weighted sum of the probability of the presence of speech in the segment according to Equation (6), where p_1 corresponds to the probability of definition of an utterance segment as speech by VAD, p_2 corresponds to the probability of definition of the utterance segment as speech by another method, and α serves as a weight coefficient. Note that the coefficient α in this regard differs from the coefficient applied in SC + OSD fusion in Section 4.4. The best configuration of the independent fusion of VAD with QE and SC with OSD for the case when more than two methods are used is denoted as

$$markup_{fusion} = \begin{cases} 1, \alpha \cdot p_1 + (1 - \alpha) \cdot p_2 > 0.5, \\ 0, \text{else.} \end{cases} \quad (6)$$

Note that the better VAD configuration is slightly different for the development and evaluation sets of AMI. The experimental results of fusion are presented in Table 6. The weight α differs depending on the methods to which fusion is applied. This allows one to estimate the best values of α, which are presented in the table as well. Based on these results, the following conclusions can be drawn:

- Using the fused VAD significantly reduces the diarization error compared the baseline VAD system. In cases of the diarization of utterances recorded in simple environments (e.g., low noise and reverberation levels), it can be assumed that VAD error will prevail over speaker errors in terms of DER.
- The QE and SC algorithms can be used for the clarification of FR errors, which leads to an error decrease in terms of DER. This effect can be observed for the high values of the fusion weight coefficient, when QE and SC help to clarify the markup in case of VAD uncertainty.
- The application of all the algorithms considered in the article in concurrence leads to a significant improvement in the quality of diarization and reduces both FR and FA errors. This effect can be observed for small α values, and probably indicates that VAD can be replaced by a fusion of the QE, SC and OSD methods.

Table 6. Results for the AMI evaluation set in terms of DER decomposed to false rejection errors, false acceptance errors and speaker errors for the fusion of different model combinations.

Fusion	DER	FR	FA	SE
VAD (baseline)	39.91–40.35–40.83	19.58	12.6	8.17
VAD + QE, $\alpha = 0.95$	28.23–28.68–29.12	4.94	15.04	8.70
VAD + SC, $\alpha = 0.58$	26.80–27.28–27.71	6.38	14.37	6.54
VAD + QE + SC + OSD, $\alpha = 0.05$	13.76–14.31–14.90	1.46	8.03	4.82

The computation of the confidence interval for the DER metric was inspired by the Python scipy.stats.bootstrap method, and consisted of the following steps:

1. Create samples of length $N - 2$ from the set of AMI scenarios, where N is the number of scenarios;
2. Compute the overall DER on that extracted subset;
3. Repeat steps 1. and 2. M times. In our case $M = 100$.

4. Compute 0.05 and 0.95 quantiles for the obtained array. The two resulting numbers correspond to the lower and upper bounds of the confidence interval.

5. Discussion

The performance of various speaker verification systems were investigated in this work with the aim of improving diarization quality. We found that the ResNet model outperformed the ECAPA-TDNN model, despite achieving worse quality in the in-domain task compared to the AMI speaker verification task. The best model configuration, diarization setup, and clustering methods were obtained for the AMI dataset to obtain better diarization quality in terms of DER. As far as the authors know, we report state-of-the-art results for the AMI dataset, reaching 1.58% DER for the AMI evaluation set for the lapel-mix microphone using a deeper ResNet model with better speaker discrimination ability. During our investigation we found that the data variability of the training dataset is important for the achievement of better diarization results, even while using 16 kHz telephone data. Further improving the quality of speaker verification models seems to be a good way to improve the quality of diarization. We confirmed the best quality of clustering for the unnormalized modification of spectral clustering compared to all other investigated methods of clustering.

In the case of the use of VAD markup, some ways to fuse different methods allowing an improvement in the quality of the estimated speech markup were considered. The fusion of VAD with methods of quality estimation, the speaker counter, and the overlapping speech detector allowed us to significantly decrease DER values from 40.35% DER to 14.31% DER. This improvement was achieved by reducing the number of errors of false rejection of speech fragments, as well as clarifying the markup of speech fragments by reducing the number of errors of false acceptance. At the same time, the number of errors involving the mixing up of speakers was still quite high, which was probably due to the difficulties in identifying mixed speech. Finding solutions for the diarization of mixed speech remains an important challenge.

By fusing the SC and OSD methods, the performance of speaker number estimation improved from 0.562 (by applying only SC) to 0.656 in terms of the F1-score. However, the estimation quality was less if the unbalanced real-life data from the AMI corpus was used for training, compared to the results obtained on synthetic mixtures of speakers. This is quite expected an explainable due to the nature of real conversations, specifically acquired from the far-field. It has been noted, though, that data augmentation which improves the balance of classes increases the estimation quality of the model. Therefore, further steps in our research include improving the fused SC and OSD system by training it on a more balanced and diverse dataset.

6. Conclusions

In this paper, various methods of audio signal processing, such as speaker number estimation, overlapping speech detection, quality estimation, and different speaker verification models, were considered with the aim of improving the quality of diarization algorithms in terms of DER. Future developments in the considered direction of research include improving the speaker verification models and training more robust methods on speech segment markup, which will allow us to further improve the quality of diarization. Furthermore, training end-to-end models for speaker diarization, employing the studied principles of speaker information estimation, may prove to be advantageous.

Author Contributions: Conceptualization, S.A. and A.G.; methodology, S.A., A.G., M.V., and A.L.; software, A.G. and A.L.; validation, S.A., A.G., and A.L.; formal analysis, S.A., A.G. and A.L.; investigation, A.G., M.V., A.L., V.Z., V.K. (Vlada Kapranova), E.T., and E.E.; resources, A.G., M.V., A.L., V.Z., V.K. (Vlada Kapranova), E.T., and E.E.; data curation, A.G., M.V., A.L., V.Z., V.K. (Vlada Kapranova), E.T., and E.E.; writing—original draft preparation, A.G., M.V., and A.L. with significant contributions from V.Z., V.K. (Vlada Kapranova), E.T., and E.E.; writing—review and editing, S.A., V.K. (Vladimir Kabarov), and Y.M.; visualization, A.G., A.L., and E.E.; supervision, V.K. (Vladimir

Kabarov) and Y.M.; project administration, S.A.; funding acquisition, S.A. All authors have read and agreed to the published version of the manuscript.

Funding: This work was partially financially supported by ITMO University.

Institutional Review Board Statement: Not applicable.

Informed Consent Statement: Not applicable.

Data Availability Statement: The data applied in this study were obtained from: the AMI corpus (https://groups.inf.ed.ac.uk/ami/corpus/ accessed on 30 May 2021); Voxceleb1 and Voxceleb2 corpora (https://www.robots.ox.ac.uk/vgg/data/voxceleb/ accessed on 15 May 2021); the NIST SRE 2019 evaluation dataset from the NIST 2019 Speaker Recognition Evaluation challenge (https://www.nist.gov/itl/iad/mig/nist-2019-speaker-recognition-evaluation accessed on 12 March 2021). Access restrictions are applied to the RusTelecom and RusIVR private Speech Technology Center corpora.

Conflicts of Interest: Authors Sergei Astapov, Aleksei Gusev, Marina Volkova, Aleksei Logunov, Valeriia Zaluskaia, Vlada Kapranova, and Yuri Matveev were employed by the company STC-innovations Ltd. The remaining authors declare that the research was conducted in the absence of any commercial or financial relationships that could be construed as a potential conflict of interest. The funders had no role in the design of the study; in the collection, analyses, or interpretation of data; in the writing of the manuscript, or in the decision to publish the results.

Abbreviations

The following abbreviations are used in this manuscript:

VAD	Voice Activity Detector
SC	Speaker Counter
OSD	Overlap Speech Detector
QE	Quality Estimation
DNN	Deep Neural Network
TDNN	Time Delay Neural Network
ECAPA-TDNN	Emphasized Channel Attention, Propagation, and Aggregation in TDNN
RIR	Room Impulse Response
RT60	Reverberation Time (the time of sound pressure reduction by 60 dB)
SNR	Signal to Noise Ratio
STFT	Short-Term Fourier transform
LMFB	Log Mel-filter Bank Energies (features)
MFCC	Mel Frequency Cepstral Coefficients (features)
CMN	Cepstral Mean Normalization
CMVN	Cepstral Mean and Variance Normalization
EER	Equal Error Rate
DER	Diarization Error Rate
SGD	Stochastic Gradient Descent
GPU	Graphics Processing Unit
DBSCAN	Density-Based Spatial Clustering of Applications with Noise
TSNE	t-Distributed Stochastic Neighbor Embedding
AM-Softmax	Additive Margin Softmax (loss)
AAM-Softmax	Additive Angular Margin Softmax (loss)
AMP	Automatic Mixed Precision

References

1. Laptev, A.; Andrusenko, A.; Podluzhny, I.; Mitrofanov, A.; Medennikov, I.; Matveev, Y. Dynamic acoustic unit augmentation with bpe-dropout for low-resource end-to-end speech recognition. *Sensors* **2021**, *21*, 3063. [CrossRef]
2. Tranter, S.E.; Reynolds, D.A. An overview of automatic speaker diarization systems. *IEEE Trans. Audio Speech Lang. Process.* **2006**, *14*, 1557–1565. [CrossRef]
3. Yella, S.H.; Bourlard, H. Overlapping Speech Detection Using Long-Term Conversational Features for Speaker Diarization in Meeting Room Conversations. *IEEE Trans. Audio Speech Lang. Process.* **2014**, *22*, 1688–1700. [CrossRef]

4. Medennikov, I.; Korenevsky, M.; Prisyach, T.; Khokhlov, Y.Y.; Korenevskaya, M.; Sorokin, I.; Timofeeva, T.; Mitrofanov, A.; Andrusenko, A.; Podluzhny, I.; et al. Target-speaker voice activity detection: A novel approach for multi-speaker diarization in a dinner party scenario. In Proceedings of the Interspeech 2020, Shanghai, China, 25–29 October 2020.
5. Dehak, N.; Kenny, P.; Dehak, R.; Dumouchel, P.; Ouellet, P. Front-end factor analysis for speaker verification. *IEEE/Trans. Audio Speech Lang. Process.* **2011**, *19*, 788–798. [CrossRef]
6. Snyder, D.; Garcia-Romero, D.; Povey, D.; Khudanpur, S. X-vectors: Robust dnn embeddings for speaker recognition. In Proceedings of the ICASSP 2018—2018 IEEE International Conference on Acoustics, Speech and Signal Processing (ICASSP), Calgary, AB, Canada, 15–20 April 2018; pp. 5329–5333.
7. Gusev, A.; Volokhov, V.; Andzhukaev, T.; Novoselov, S.; Lavrentyeva, G.; Volkova, M.; Gazizullina, A.; Shulipa, A.; Gorlanov, A.; Avdeeva, A.; et al. Deep Speaker Embeddings for Far-Field Speaker Recognition on Short Utterances. *Odyssey 2020*, *2020*, 179–186. [CrossRef]
8. Lavechin, M.; Gill, M.P.; Bousbib, R.; Bredin, H.; Garcia-Perera, L.P. End-to-end Domain-Adversarial Voice Activity Detection. *arXiv* **2019**, arXiv:1910.10655.
9. Lavrentyeva, G.; Volkova, M.; Avdeeva, A.; Novoselov, S.; Gorlanov, A.; Andzhukaev, T.; Ivanov, A.; Kozlov, A. Blind Speech Signal Quality Estimation for Speaker Verification Systems. In Proceedings of the Interspeech 2020, Shanghai, China, 25–29 October 2020; pp. 1535–1539. [CrossRef]
10. AMI Corpus. Available online: https://groups.inf.ed.ac.uk/ami/corpus/ (accessed on 30 May 2021).
11. Andrei, V.; Cucu, H.; Burileanu, C. Overlapped Speech Detection and Competing Speaker Counting—Humans Versus Deep Learning. *IEEE J. Sel. Top. Signal Process.* **2019**, *13*, 850–862. [CrossRef]
12. Chung, J.S.; Lee, B.J.; Han, I. Who said that?: Audio-visual speaker diarisation of real-world meetings. *arXiv* **2019**, arXiv:1906.10042.
13. Desplanques, B.; Thienpondt, J.; Demuynck, K. ECAPA-TDNN: Emphasized Channel Attention, Propagation and Aggregation in TDNN Based Speaker Verification. *arXiv* **2020**, arXiv:2005.07143v3.
14. Kumar, A.K.; Waldekar, S.; Saha, G.; Sahidullah, M. Domain-Dependent Speaker Diarization for the Third DIHARD Challenge. *arXiv* **2021**, arXiv:2101.09884.
15. Ferrés, M.; Bourlard, H. Speaker diarization and linking of large corpora. In Proceedings of the 2012 IEEE Spoken Language Technology Workshop (SLT), Miami, FL, USA, 2–5 December 2012; pp. 280–285.
16. Cristia, A.; Ganesh, S.; Casillas, M.; Ganapathy, S. Talker diarization in the wild: The case of child-centered daylong audio-recordings. In Proceedings of the Interspeech 2018, Hyderabad, India, 2–6 September 2018; pp. 2583–2587.
17. von Luxburg, U. A Tutorial on Spectral Clustering. *arXiv* **2007**, arXiv:0711.0189.
18. Bissig, P.; Foerster, K.; Tanner, S.; Wattenhofer, R. Distributed discussion diarisation. In Proceedings of the 2017 14th IEEE Annual Consumer Communications I & Networking Conference (CCNC), Las Vegas, NV, USA, 8–11 January 2017; pp. 1032–1037.
19. SileroTeam. Silero VAD: Pre-Trained Enterprise-Grade Voice Activity Detector (VAD), Number Detector and Language Classifier. 2021. Available online: https://github.com/snakers4/silero-vad (accessed on 29 June 2021).
20. Povey, D.; Ghoshal, A.; Boulianne, G.; Burget, L.; Glembek, O.; Goel, N.; Hannemann, M.; Motlicek, P.; Qian, Y.; Schwarz, P.; et al. The Kaldi speech recognition toolkit. In Proceedings of the IEEE 2011 Workshop on Automatic Speech Recognition and Understanding, Waikoloa, HI, USA, 11–15 December 2011.
21. Dawalatabad, N.; Ravanelli, M.; Grondin, F.; Thienpondt, J.; Desplanques, B.; Na, H. ECAPA-TDNN Embeddings for Speaker Diarization. *arXiv* **2021**, arXiv:2104.01466.
22. Landini, F.; Profant, J.; Diez, M.; Burget, L. Bayesian HMM clustering of x-vector sequences (VBx) in speaker diarization: Theory, implementation and analysis on standard tasks. *Comput. Speech Lang.* **2022**, *71*, 101254. [CrossRef]
23. Nagrani, A.; Chung, J.S.; Zisserman, A. VoxCeleb: A large-scale speaker identification dataset. In Proceedings of the Interspeech 2017, Stockholm, Sweden, 20–24 August 2017.
24. Chung, J.S.; Nagrani, A.; Zisserman, A. VoxCeleb2: Deep Speaker Recognition. In Proceedings of the Interspeech 2018, Hyderabad, India, 2–6 September 2018.
25. Nagrani, A.; Chung, J.S.; Xie, W.; Zisserman, A. Voxceleb: Large-scale Speaker Verification in the Wild. *Comput. Speech Lang.* **2020**, *60*, 101027. [CrossRef]
26. Sadjadi, S.O.; Greenberg, C.; Singer, E.; Reynolds, D.; Mason, L.; Hernandez-Cordero, J. The 2019 NIST Speaker Recognition Evaluation CTS Challenge. In Proceedings of the Odyssey 2020 The Speaker and Language Recognition Workshop, Tokyo, Japan, 1–5 November 2020; pp. 266–272. [CrossRef]
27. Ronneberger, O.; Fischer, P.; Brox, T. U-Net: Convolutional Networks for Biomedical Image Segmentation. *arXiv* **2015**, arXiv:1505.04597.
28. Shruti, J. A survey of loss functions for semantic segmentation. In Proceedings of the 2020 IEEE Conference on Computational Intelligence in Bioinformatics and Computational Biology (CIBCB), Via del Mar, Chile, 27–29 October 2020.
29. Brümmer, N.; De Villiers, E. The speaker partitioning problem. In Proceedings of the Odyssey 2010, the Speaker and Language Recognition Workshop, Brno, Czech Republic, 28 June–1 July 2010; p. 34.
30. Stöter, F.R.; Chakrabarty, S.; Edler, B.; Habets, E.A. Classification vs. regression in supervised learning for single channel speaker count estimation. In Proceedings of the 2018 IEEE International Conference on Acoustics, Speech and Signal Processing (ICASSP), Calgary, AB, Canada, 15–20 April 2018; pp. 436–440.

31. Cornell, S.; Omologo, M.; Squartini, S.; Vincent, E. Detecting and counting overlapping speakers in distant speech scenarios. In Proceedings of the Interspeech 2020, Shanghai, China, 25–29 October 2020.
32. Yousefi, M.; Hansen, J. Real-time Speaker counting in a cocktail party scenario using Attention-guided Convolutional Neural Network. *arXiv* **2021**, arXiv:2111.00316.
33. Stöter, F.R.; Chakrabarty, S.; Edler, B.; Habets, E.A. CountNet: Estimating the number of concurrent speakers using supervised learning. *IEEE/ACM Trans. Audio Speech Lang. Process.* **2018**, *27*, 268–282. [CrossRef]
34. Timofeeva, E.; Evseeva, E.; Zaluskaia, V.; Kapranova, V.; Astapov, S.; Kabarov, V. Improvement of Speaker Number Estimation by Applying an Overlapped Speech Detector. In *Speech and Computer*; Karpov, A.; Potapova, R., Eds.; Springer International Publishing: Cham, Switzerland, 2021; pp. 692–703.
35. Bredin, H.; Yin, R.; Coria, J.M.; Gelly, G.; Korshunov, P.; Lavechin, M.; Fustes, D.; Titeux, H.; Bouaziz, W.; Gill, M.P. Pyannote. audio: Neural building blocks for speaker diarization. In Proceedings of the ICASSP 2020—2020 IEEE International Conference on Acoustics, Speech and Signal Processing (ICASSP), Barcelona, Spain, 4–9 May 2020; pp. 7124–7128.
36. Bullock, L.; Bredin, H.; Garcia-Perera, L.P. Overlap-aware diarization: Resegmentation using neural end-to-end overlapped speech detection. In Proceedings of the ICASSP 2020—2020 IEEE International Conference on Acoustics, Speech and Signal Processing (ICASSP), Barcelona, Spain, 4–9 May 2020; pp. 7114–7118.
37. Kunešová, M.; Hrúz, M.; Zajíc, Z.; Radová, V. Detection of overlapping speech for the purposes of speaker diarization. In Proceedings of the International Conference on Speech and Computer, Istanbul, Turkey, 20–25 August 2019; pp. 247–257.
38. Otterson, S.; Ostendorf, M. Efficient use of overlap information in speaker diarization. In Proceedings of the 2007 IEEE Workshop on Automatic Speech Recognition & Understanding (ASRU), Kyoto, Japan, 9–13 December 2007; pp. 683–686.
39. Boakye, K.; Trueba-Hornero, B.; Vinyals, O.; Friedland, G. Overlapped speech detection for improved speaker diarization in multiparty meetings. In Proceedings of the 2008 IEEE International Conference on Acoustics, Speech and Signal Processing, Las Vegas, NV, USA, 30 March–4 April 2008; pp. 4353–4356.
40. Alexeev, A.; Kukharev, G.; Matveev, Y.; Matveev, A. A highly efficient neural network solution for automated detection of pointer meters with different analog scales operating in different conditions. *Mathematics* **2020**, *8*, 1104. [CrossRef]
41. Park, D.S.; Chan, W.; Zhang, Y.; Chiu, C.C.; Zoph, B.; Cubuk, E.D.; Le, Q.V. SpecAugment: A Simple Data Augmentation Method for Automatic Speech Recognition. In Proceedings of the Interspeech 2019, Graz, Austria, 15–19 September 2019; pp. 2613–2617. [CrossRef]
42. Ravanelli, M.; Parcollet, T.; Plantinga, P.; Rouhe, A.; Cornell, S.; Lugosch, L.; Subakan, C.; Dawalatabad, N.; Heba, A.; Zhong, J.; et al. SpeechBrain: A General-Purpose Speech Toolkit. *arXiv* **2021**, arXiv:2106.04624.
43. Brock, A.; De, S.; Smith, S.L.; Simonyan, K. High-Performance Large-Scale Image Recognition Without Normalization. *arXiv* **2021**, arXiv:2102.06171.
44. Huang, Y.; Wang, Y.; Tai, Y.; Liu, X.; Shen, P.; Li, S.; Li, J.; Huang, F. CurricularFace: Adaptive Curriculum Learning Loss for Deep Face Recognition. *arXiv* **2020**, arXiv:2004.00288.
45. Gusev, A.; Vinogradova, A.; Novoselov, S.; Astapov, S. SdSVC Challenge 2021: Tips and Tricks to Boost the Short-Duration Speaker Verification System Performance. In Proceedings of the Interspeech 2021, Brno, Czech Republic, 30 August–3 September 2021. [CrossRef]
46. van Leeuwen, D.; Brümmer, N. An Introduction to Application-Independent Evaluation of Speaker Recognition Systems. In *Speaker Classification I. Lecture Notes in Computer Science*; Springer: Berlin/Heidelberg, Germany, 2007; Volume 4343._19. [CrossRef]
47. Haasnoot, E.; Khodabakhsh, A.; Zeinstra, C.; Spreeuwers, L.; Veldhuis, R. FEERCI: A Package for Fast Non-Parametric Confidence Intervals for Equal Error Rates in Amortized O(m log n). In Proceedings of the 2018 International Conference of the Biometrics Special Interest Group (BIOSIG), Darmstadt, Germany, 26–28 September 2018; pp. 1–5. [CrossRef]
48. Biagetti, G.; Crippa, P.; Falaschetti, L.; Orcioni, S.; Turchetti, C. Robust Speaker Identification in a Meeting with Short Audio Segments. In *Intelligent Decision Technologies 2016*; Czarnowski, I., Caballero, A.M., Howlett, R.J., Jain, L.C., Eds.; Springer International Publishing: Cham, Switzerland, 2016; pp. 465–477.

Article

Tags' Recommender to Classify Architectural Knowledge Applying Language Models

Gilberto Borrego [1,†], Samuel González-López [2,†] and Ramón R. Palacio [3,*,†]

1. Departamento de Computación y Diseño, Instituto Tecnológico de Sonora, Ciudad Obregón 85000, Mexico; gilberto.borrego@itson.edu.mx
2. Department of Information Technologies, Universidad Tecnológica de Nogales, Nogales 84097, Mexico; sgonzalez@utnogales.edu.mx
3. Unidad Navojoa, Instituto Tecnológico de Sonora, Navojoa 85860, Mexico
* Correspondence: ramon.palacio@itson.edu.mx
† These authors contributed equally to this work.

Abstract: Agile global software engineering challenges architectural knowledge (AK) management since face-to-face interactions are preferred over comprehensive documentation, which causes AK loss over time. The AK condensation concept was proposed to reduce AK losing, using the AK shared through unstructured electronic media. A crucial part of this concept is a classification mechanism to ease AK recovery in the future. We developed a Slack complement as a classification mechanism based on social tagging, which recommends tags according to a chat/message topic, using natural language processing (NLP) techniques. We evaluated two tagging modes: NLP-assisted versus alphabetical auto-completion, in terms of correctness and time to select a tag. Fifty-two participants used the complement emulating an agile and global scenario and gave us their complement's perceptions about usefulness, ease of use, and work integration. Messages tagged through NLP recommendations showed fewer semantic errors, and participants spent less time selecting a tag. They perceived the component as very usable, useful, and easy to be integrated into the daily work. These results indicated that a tag recommendation system is necessary to classify the shared AK accurately and quickly. We will improve the NLP techniques to evaluate AK condensation in a long-term test as future work.

Keywords: agile global software engineering; architectural knowledge management; natural language processing; knowledge condensing

1. Introduction

Architectural knowledge (AK) is a core part of any software project [1], and it is commonly documented based upon standards [2]. This documentation eases the knowledge management cycle [3] (create/capture, share/disseminate, acquire/apply), even when companies practice global software engineering (GSE). In GSE, transferring/sharing documents among the distributed facilities is commonly done to decrease the effect of the four distances in this paradigm [4,5] (physical, temporal, linguistic, and cultural).

Nowadays, GSE companies are adopting agile software development, which caused the arising of the agile GSE approach; in fact, VersionOne (https://explore.digital.ai/state-of-agile/14th-annual-state-of-agile-report, accessed on 15 March 2021) reports in 2020 that 71% of surveyed agile enterprises practice agile GSE. This trend causes a paradigms contradiction referring to knowledge managing. The agile paradigm states that functional software and face-to-face communication are preferred over processes, and comprehensive documentation [6]. In practice, most agile projects' documentation could be replaced by enhancing informal communication since the agile paradigm proposes a stronger emphasis on tacit knowledge rather than explicit knowledge [7]. This emphasis on tacit knowledge has to be done without disregarding formal documentation [8]; however, in the agile GSE practice, this implies a reduction of AK documentation or even its disappearance [8–10], known as documentation debt [11]. This situation affects the AK management since most of the projects' AK remains tacit [12], hindering the knowledge management cycle in agile GSE environments because traditional knowledge managing on GSE is based on explicit knowledge sharing.

The prevalence of AK documentation debt and the affectation on the knowledge management cycle eventually cause AK vaporization [13]. It could trigger some of the following problems in agile GSE projects [14–16]: poorly understood technical solutions, reduced knowledge transfer between teams, low quality, defects in software evolution and maintenance, difficulty in conducting tests, stress with stakeholders, and time wasted by experts answering questions and attempting to find solutions to problems previously solved. It is worth highlighting that constant questions could cause interpersonal relationship erosion, which could affect the knowledge flow [17], and trust relationship building, which is essential for agile teams.

In the aim to address the above problems, in a previous study [18] we presented the knowledge condensation concept, which takes advantage of the agile GSE developers' preference to share AK by unstructured textual and electronic media (UTEM), e.g., instant messengers, emails, etc., [10,19]. The knowledge condensation concept includes a knowledge classification mechanism, which was initially based on social tagging to tag relevant AK during their interactions. These tags have to be registered by the development teams to relate to the teams' jargon. Furthermore, these tags must be related to a set of meta-tags, which are semantic anchors to ease the AK later retrieval in case the exact name of a tag was forgotten. This semi-fixed tagging mechanism could avoid the free and unassisted tagging problems. These problems are tag explosion, interpretation differences of a tag's meaning, incomplete context to understand a tag; tags that only make sense when used together (known as composite tags), and tags with the same meaning but written differently (known as obscure similarity) [20]. These problems could lead to tagging difficulties which could ruin the AK classification mechanism, and consequently, they would cause information retrieval problems [21]. To preserve the classification mechanism, in the same study [18], we presented a tagging helper implemented for Skype, which auto-completes tags in an alphabetic way during developers' UTEM interactions, helping them to select an existing tag and to type it correctly. Both AK classification mechanism and tagging service were evaluated, and we obtained broad acceptance by the participants. However, they suggested including a more intelligent way to help the tagging action. We considered it essential to increase the knowledge condensation concept adoption in a real agile GSE environment.

Therefore, the tagging service must minimize the negative cost that interruption could cause [22,23], since developers must redirect their attention from the main activity to properly tag the current interaction and then return naturally to its primary activity. Thus, a tagging service is required to ease the knowledge flow and not become a constant source of disruption [24,25].

Based on the stated above, we developed an AK classification mechanism based on semi-fixed tagging, which was implemented as a Slack (https://slack.com/, accessed on 20 October 2021) service, including a tags recommendation module based on natural language processing (NLP) and statistical language models techniques.

This paper aims to determine with which tagging method, either assisted by auto-completion or by NLP, users select tags with a better coherence (named in this paper as semantic correctness) regarding the context of the messages. A coherent tagging is essential to the AK condensation concept since it is the base to find the required AK in the future. Another objective in this paper is to determine with which tagging method a developer could select in a faster way a given tag. The time to tag a message is also an essential factor to consider since interruptions could cause negative consequences in the developer work (as we stated before), as well as the time to accomplish a task is relevant to enhance the usability of any software product [26]. Thus, a quick tag selection could increase the adoption possibility of our proposed classification mechanism, as well as the knowledge condensation concept per se. Thus, the research questions of this paper are the following: (RQ1) With which tagging method (assisted by auto-completion or assisted by NLP) does the tag selection by users fit better to the message context?; (RQ2) With which tagging method (assisted by auto-completion or assisted by NLP) the tag selection by users is faster? and (RQ3) Which is the perception of usefulness, ease of use, and integration into the work of the tagging service implemented in Slack?

The main contributions of this paper are the following: (1) the implementation of a tagging service based on NLP to help to condense AK; (2) evidence about an NLP tagging service cause that users tag faster and more coherently than using tag auto-completion; (3) encouraging evaluation results of the implemented tagging service in terms of usefulness, ease of use and integration in software developer activities; (4) relevant results about the pertinence of the recommended tags by the NLP mechanism; and (5) a corpus of 291 tagged messages contained in seven categories.

The remainder of this paper is organized as follows: Section 2 presents the work related to AK extraction/classification and tagging recommendation in software engineering; in Section 3, we present the Slack service used to implement the AK classification mechanism; Section 4 presents a description of the NLP models used to obtain the tags recommendation; Section 5 presents the method used to evaluate the service; in Section 6 and Section 7, we, respectively, present the results of the evaluation and its threats of validity. Finally, in Section 8 we discuss the obtained results and present our conclusions in Section 9.

2. Related Work

In this section, we present the related work about AK, where we clarify the abstractions levels used in this concept and how AK management is a challenge in agile GSE nowadays. Furthermore, we present the approaches to address the lack of explicit AK in agile GSE, based on using repositories, tagging, Q&A sites, to leverage the work done by software developers to provide a low-intrusive means of capturing AK during day-to-day work.

2.1. Architectural Knowledge and Agile Global Software Engineering

According to Kruchten et al. [27] AK is composed of architecture design and the design decisions and rationale used to attain architectural solutions. This definition could be clear for everyone; however, there are many definitions of the concept of software architecture; thus, the term architecture design is not as straightforward as possible. The definitions of software architecture from Bass et al. [28], from the ISO/IEC/IEEE 42010:2011 standard [29], from Kruchten [30] and many others, could build up the idea that software architecture is a very abstract and general definition of a structure on which software will be developed. As architecture implies abstraction, it suppresses purely local information; thus, private details of components are not architectural [31].

Deeping in the software architecture concepts, there also exists the architectural views definition: a representation of a coherent set of architectural elements as written by and read by system stakeholders [28]. Thus, depending on the stakeholders, the level of design detail could vary. The typical architectural views include the functional view (called the logical view), the concurrency view (called the process or thread view), the code view, the development view (called the implementation view), and the physical view [31]. The means above that detailed views on code and development could be considered architectural topics. Further, if the concern's scope is limited to a subsystem within a system, what is considered architectural is different from what the software architect considers architectural [31]. Therefore, the context influences what is architectural, leading to the concepts of macro-architecture and micro-architecture. Macro-architecture covers the architecture levels to which architecturally relevant elements are assigned (large-scale architecture); thus, it covers aspects such as requirements, decisions, and structures at a high level of abstraction: decisions concerning necessary system interfaces. Micro-architecture covers aspects with a lower level of abstraction, i.e., detailed design or "small-scale" architecture, which is closely associated with the source code with no fundamental influence on an architecture [32].

Then, the concept of AK (design decisions and rationale used to attain an architectural design) takes another dimension. Software architecture could cover either high abstract and general topics (macro-architecture) or detailed design at low abstraction levels (micro-architecture) regarding the architectural concern's scope.

On the other hand, AK management is challenging in agile GSE since knowledge management is a challenge too [33–35]. Knowledge capturing is a critical phase of the AK management process in agile GSE environment [36]; however, the agile developers' attitudes [37] towards this phase cause documentation debt [11]. Since an agile GSE environment leads to a lack of captured AK, the knowledge management phases of sharing/disseminating and acquiring/applying are also affected because AK is shared and acquired based on inappropriate documentation or even tacit knowledge. Souza et al. [38] determined that current architectural derivation methods are heavily based on the architects' tacit knowledge, and this does not offer sufficient support for inexperienced architects or novice developers. The limitations of the current derivations methods are the following: undetailed decision-making process, inconsistency between the artifacts (especially those referring to micro-architecture), and semantic loss, and lack of traceability between models (between/within macro-architecture and micro-architecture).

2.2. Approaches to Address the Lack of Explicit Architectural Knowledge

One of the approaches to address the lack of explicit AK at the micro-architecture level is mining issue trackers, version control repositories (generally from open source software projects), or any other document which could contain AK. The main objective is to extract large sets of architectural design concepts [39], or automatically recover design decisions from projects' artifacts [40], which can be textual or based on unified modeling language [41]. However, none of these works consider that in agile environments exist documentation debt; thus, the mining sources used in these works could not exist o could be incomplete or outdated. Furthermore, mining repositories could be inefficient on small projects with a non-rigid discipline regarding the commits and bug fixing comments. Further, challenges associated with the volume, velocity, and variety of the data set and other challenges about extensive data systems were identified.

Another proposal to address the lack of explicit AK is using tags recommendation systems to classify work items during the development cycle. Tagging has been used in software engineering for information finding, team coordination, and cooperation, and helping to bridge socio-technical issues [42,43]. Automatic tag recommendation in software engineering started with the TAGREC method [44] to recommend tags for work items in IBM Jazz, based on the fuzzy set theory. More recently, neural networks had been used to recommend tags for question and answers (Q&A) sites [45], finding that this technique could infer the context of the posts and consequently it could recommend accurate tags for new posts. On the same line, Mushtaq et al. [46] proposed a framework to obtain explicit and implicit knowledge from community-based Q&A sites (e.g., StackOverflow) through mining techniques to support AK management; however, the authors did not report any implementation. In the Parra et al. work [47], a process to extract relevant keywords from video tutorials was proposed, which uses the video title, description, audio transcription, as well as external resources related to the video content, such as StackOverflow publications and its tags. This method showed that it could automatically extract tags that encompass up to 66% of the information developers consider relevant to outline the content of a software development video.

Another way of tagging is presented by Jovanovic et al. [48], called semantic tagging, which uses Wikipedia information to establish the tagging process semantically and to provide a methodical approach to apply tagging in software engineering content. In the same sense, TRG [49] proposes to discover and enrich tags focused on open source software through using a semantic graph to model the semantic correlations between the tags and words of the software descriptions. A similar approach was used by Alqahtani and Rilling [50] who proposed an ontology-based software security tagger framework to automatically identify and classify cybersecurity-related entities and concepts in the text of software artifacts.

Most of the above-cited works use the traditional approaches in tags recommendation (EnTagRec, TagMulRec, and FastTagRec). They were compared against deep learning-based approaches (TagCNN, TagRNN, TagHAN, and TagRCNN) [51], obtaining that the performance of TagCNN and TagRCNN was more suitable in terms of performance than the traditional approaches. However, all these approaches do not consider the "cold start" problem [52], i.e., the absence of previous information about the target object.

The presented works evidence that Q&A sites, wikis, or any other similar platform can be considered as a source of AK, as was stated by Soliman et al. [53], who analyzed and classified StackOverflow posts, and these were contrasted with software engineers. They obtained that there is valuable AK particularly related to technological decisions. However, these discussions do not include specific project AK such as business rules, architectural decisions based on particular conditions of the project, etc. Therefore, organizations need to unify their AK sources, to have a similar resource as Q&A systems. CAKI [54] is a proposed solution to the necessity of internal AK, which works as a continuous integration of organization-internal and external AK sources, with enhanced semantic reasoning and personalized capabilities.

There are different approaches to enhance AK managing, focused on the extraction, classification, and searching of knowledge. Many efforts are focused on extracting/classifying AK from public Q&A services sites which concentrate enormous volumes of information, experiences, and knowledge related to technical aspects of software engineering, using different artificial intelligence techniques and data mining. However, AK is not stored in those Q&A services, such as discussions about project business rules, architectural decisions that depend on the project conditions, and even AK related to the project deployment. This AK is usually stored in the UTEM logs of the software company [9], and these information sources are our focus with the knowledge condensation concept. To help classify the unstructured AK in the UTEM logs, we chose assisted tagging to recommend adequate tags to project stakeholders. In the following section, we describe the Slack service developed to recommend tags using NLP models.

3. Architectural Knowledge Condensation and the Slack Tagging Service

The AK condensation concept was proposed in a previous work [18] as a response to reduce the AK vaporization in agile GSD, taking advantage of the preference of sharing AK through UTEM in these kinds of environments [9,10]. To instantiate this concept, we have to ensure the following items [18]:

- Accessible UTEM logs. All the stakeholders involved in an agile GSE project must access the information contained in the UTEM log files to consult the AK that is shared among the development team.
- UTEM log classification mechanism to structure the shared information to ease the AK retrieval. This mechanism must be based on a semantic scheme representing the AK shared through UTEM.
- AK searching mechanism. All the stakeholders could use the semantic scheme to find valuable AK with less effort in the structured UTEM logs.

In the aim to instantiate the AK condensation concept we chose Slack, an application that eases communication and collaboration within teams, which is booming in agile development, and researchers have even studied how developers use Slack [55–57]. Slack can be considered a UTEM since the messages shared between the interested parties have no fixed structure to ease the retrieval of AK. Given the popularity of Slack, we decided to develop a Slack tagging service so that the concept of knowledge condensation could eventually be evaluated in a real agile GSE environment. While Slack uses social tagging as part of its functionality, and there are some add-ons based on the use of tags, these present the problems associated with free tagging: tag explosion, interpretation differences of a tag's meaning, incomplete context to understand a tag, etc. [20]. Furthermore, there are no assistants that implement intelligent elements to suggest tags.

Hereunder, we describe the operation of the service in terms of the elements of the concept of knowledge condensation [18], excepting the element named "Architectural knowledge searching mechanism", since it is out of the scope of this study (an example of the implementation of this mechanism can be seen in our past work [18]). In the last part of this section, we present a scenario to explain how the AK condensation concept could work in an agile GSE environment and the potential benefits of its implementation.

3.1. Accessible Messages of the UTEM Log Files

We developed a component in Node.js, which extracts the messages periodically (or at request) using the Slack API and then stores the messages in a common repository (Algolia (https://www.algolia.com/, accessed on 29 October 2021). The main objective is to keep the Slack messages (previously classified by tags—see next section) and another UTEM's messages located in one single point so that the AK can easily be found.

3.2. UTEM Log Files Classification Mechanism

In a previous work [18], we learned that tagging is feasible to classify the AK stored in the UTEM log files. With this focus, we proposed that developers tag the UTEM messages they consider relevant in terms of AK during their daily interactions to recover them in the future. To avoid tags explosion phenomena [20] (occurring when free tagging is allowed), we chose a semi-fixed tagging in a past work [18]. This type of tagging has a meta-tags base model (obtained from a previous study [9], see Figure 1), which represents the AK shared in UTEM in an agile GSE environment. Each meta-tag is related to user-tags which developers register in a web application at the beginning of each agile development cycle after they agree about which tags they will use during that cycle. The meta-tag model could be changed by another model that fits in other environments, such as the work of Martinez et al. [58] who formalize that model as an ontology for general use in software projects. The main objective of this model is to have a semantic anchor to recover AK in the future.

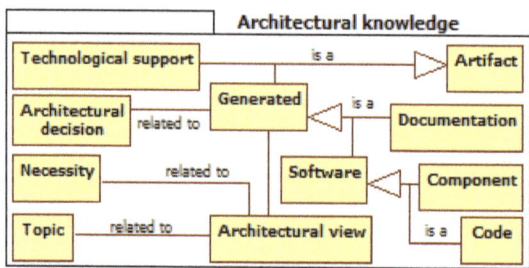

Figure 1. Meta-tags model, representing the architectural knowledge shared in agile GSE environments (adapted from [9]).

To ease the tags recalling during the interactions through Slack, we implemented a tagging system on which developers have to select the option "Tag! Tagger" from the options that appear for each message. Then, a dialogue box appears to allow the selection from a tags' list previously recorded, including meta-tags (see Figure 2A). This selection list includes an auto-completion feature that filters the available tags as the developer writes. It is also possible to enter customized (free) tags in a text field. In the final selection list, developers can view the tags recommendation list, which comes from the language models (see Figure 2B). On this list, the elements are the suggested tags. These elements are ordered regarding how semantically close the tag is to the message to be tagged. It is worth remarking that the tagging system has three different ways to tag a message only for evaluation purposes.

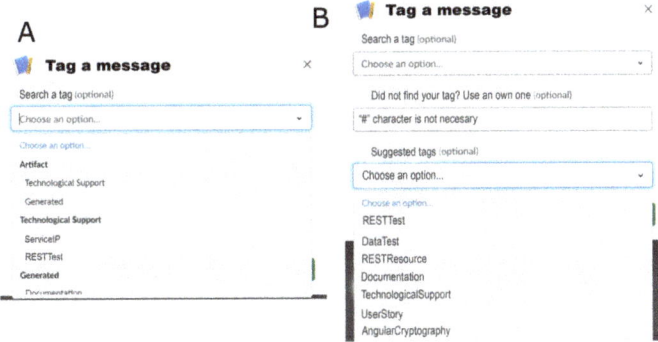

Figure 2. Graphic user interfaces of Slack tagging service. (**A**) Auto-completion assistance to select a tag; (**B**) tags recommendation list of the natural language processing (NLP) assistance comes from the language models. The middle textbox is used to type tags freely (only for evaluation purposes).

3.3. Scenario of Architectural Knowledge Condensation

This part exemplifies how developers could use the Slack tagging service in an agile GSE scenario where the knowledge condensation concept is implemented.

First, developers (local and remote) during a kick-off meeting of a development cycle (e.g., Scrum sprint meeting) agree on the tags to use during the next cycle depending on the topics covered by the user stories to be developed. Next, developers register the agreed tags in the tag management Web application (see Figure 3B), and the conventional flow of an agile methodology continues.

In an agile environment, it is common that developers neglect AK documentation activities, as working software and individuals' interaction are preferred over comprehensive documentation, process, and tools [6]. In addition, there generally exists time pressures in agile GSE environments [36], and developers show a lousy attitude towards documentation [37]. These situations generate documentation debt [11], which is addressed by constantly interacting between remote and local team members via UTEM to acquire and share AK. During those interactions, developers can tag the AK they feel necessary to retrieve later (see Figure 3B), supported by the tagging service for Slack or another UTEM, such as email, Trello, Skype (see our past implementation in [18]). The tagged messages are stored in the respective UTEM logs, which the gathering service frequently reads and sends the messages to a repository (see Figure 3A).

Finally, due to the documentation debt, it is frequent that developers do not remember some architectural details (e.g., technical specifications, detailed design, application deployment issues, etc.) either during the same development cycle or in a later one [9]. It is even common that they do not remember the exact date or the communication media used to share AK with their counterparts. However, since team members tagged their interactions and these are located in a common repository, the project AK could be retrieved using the AK search engine presented in [18], which is out of the scope of this paper (see Figure 3C).

Although we describe this scenario sequentially, the technology elements involved (tagging assistants, tag management web application, gathering service, AK search engine, etc.) can come into play at any time. For example, tag registration and/or deletion could be performed at any time during the development cycle, as long as there is agreement among team members. Furthermore, although tagging is fundamental to this implementation of knowledge condensation, knowledge retrieval can be performed even before any message is tagged, either by retrieving by period, author, recipient, etc. However, not tagging conversations can lead to the loss of semantics and context of knowledge over time since tags and the meta-tag structure to which they are related provide a semantic anchor that would offer a way to remember or intuit the context of the conversation being consulted.

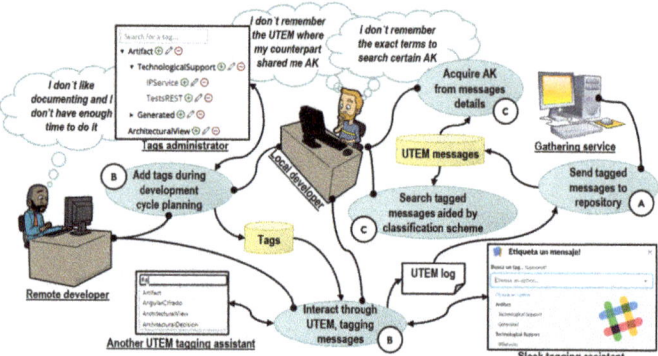

Figure 3. Rich picture of an AK condensation scenario based on [18]. It includes activities (ovals) corresponding to the three elements of the AK condensation concept, A = Accessible UTEM logs information, B = UTEM logs Classification mechanism, C = AK searching mechanism. Bulleted lines represent links between activities and who performs them. Arrowed lines represent links between activities and artifacts. There are three types of artifacts: resulting artifacts (activities' outgoing arrows), source artifacts (activities' in-going arrows), and interacting artifacts (linked with double arrow lines).

3.4. Potential Benefits of the Architectural Knowledge Condensation

Listed below are the potential benefits of fully implementing the architectural knowledge condensation concept.

- Reduction of interruptions. Since full implementation of this concept provides a search engine for AK, a team member who has a question about any AK topic could first use the search engine before interrupting one of their or her teammates to ask a question.
- Reduced time to find AK. AK condensation offers a common point to search for shared knowledge in different UTEM, in addition to offering search filters that most UTEM do not offer, such as search by period, by author, by the recipient, by UTEM source, and by tag.
- Reduction of development tasks time. As a consequence of the two previous points, it is expected that the time to complete development tasks will be reduced since the project's AK will be more accessible through the AK search engine.

A crucial point to obtain the benefits listed above is that the AK classification mechanism is efficient and adapted to the dynamics of an agile environment to ensure the adoption of the technological elements that implement such mechanisms. That is why in this paper, we focus on evaluating how well an NLP tagging service suggests tags according to the context, intending to ease the interactions tagging. Furthermore, it is of our interest to evaluate how quickly the tagging action can be performed since the time to accomplish a tag is an essential factor to the future adoption of any software tool [26].

4. Language Models Development and Integration to Slack Tagging Service

As we mentioned before, we incorporated NLP techniques and statistical language models into the Slack tagging service. Language models use probabilities for their operation and specifically assign a value to a sequence of words [59]. An n-gram is a sequence of n terms, which can be words or characters, e.g., "story user" is a sequence of two words and is called 2-gram or bigram. In this way, we can have different configurations of n-grams. N-gram models can compute the probability of a sequence of words (text segment) by estimating each word's probability in an n-gram given the previous terms. Equation (1) computes the probabilities for any string of terms.

$$P(w_{1:n}) = P(w_1)P(w_2|w_1)P(w_3|w_{1:2})\ldots P(w_n|w_{1:n-1})$$

$$P(w_{1:n}) \approx \prod_{i=1}^{n} P(w_n|w_{1:n-1}) \quad (1)$$

In the aim of integrating the use language models to suggest tags during Slack interactions, we executed the following phases: (1) corpus development, (2) language models development, and (3) language models and Slack tagging service integration.

4.1. Corpus Development

Language models require initial data (corpus) to be developed; thus, we took the UTEM logs generated in our previous study [18], in which participants interacted through Skype following a script. In the referenced study, participants tagged Skype messages following the script indications, and then, we determined which messages were tagged coherently. It is important to highlight that participants used the same set of tags used in this study, and those tags were related to the meta-tags model presented in the previous section (see Figure 1).

As the next step, we selected the coherently tagged messages to determine which tags were used more frequently. We obtained 291 messages tagged with the following tags (Table 1 the tags distribution): Encryption, Documentation, UserStory, RESTTest, RESTResource, TechnologicalSupport and TestData. It is important to point out that Documentation and TechnologicalSupport are meta-tags. Thus, we had the base to develop seven language models, one per each frequent tag with this data. We used this corpus to train the language models generated by each category (tags).

Table 1. Distribution of the 291 tagged messages selected from our previous study [18].

Tag/Meta-Tag	Number of Messages
UserStory	88
RESTResource	61
RESTTest	43
Documentation	36
Encryption	26
TestData	23
TechnologicalSupport	16

4.2. Language Models Development

The developed method seeks to capture features of the seven tags selected from the original corpus. As the input of our language model, we used the lemma and the grammatical class. In the lemmatization process, the words' inflections are removed, and the words' root is obtained. For example, the word "computing" has the word "compute" as its lemma. On the other hand, the grammatical categories are the different classifications in which words are grouped, for example, nouns, articles, adjectives, verbs, pronouns, prepositions, conjunctions, and interjections. For example, the original comment from a developer in the #UserStory category "the resource you consult is the same as story 2" provides for the word story "story NN". The NN grammatical class represents a noun. The probability of #UserStory example as bigrams could be computed with the Equation (2). This probability is affected by the appearance of each term in the entire collection of messages (corpus).

$$P(W_n|W_{n-1}) = \frac{C(W_{n-1}W_n)}{C(W_{n-1})} \quad (2)$$

The Equation (2) is called maximum likelihood estimation or MLE. The MLE estimates for an n-gram model's parameters by getting counts from a corpus and normalizing the counts; therefore, the values lie between 0 and 1.

Let us see an example using bigrams with corpus of three sentences:

1. *B* them. I am programming *END*
2. *B* programming I am *END*
3. *B* I do not like to design *END*

Below we show the result for some bigrams of this short corpus:

$$P(I \mid *B*) = \frac{2}{3} = 0.67,$$

the word I appears two times after *B* and *ENDS* appears three times.

$$P(*END* \mid programming) = \frac{1}{2} = 0.5$$
$$P(programming \mid *B*) = \frac{1}{3} = 0.33$$
$$P(programming \mid am) = \frac{1}{2} = 0.5$$
$$P(am \mid I) = \frac{2}{3} = 0.67$$

To compute the probability of the first sentence, we would have the following expression:

$$\begin{aligned} P(S3) &= P(I \mid *B*) + P(am \mid I) + P(programming \mid am) \\ &= +P(*END* \mid programming) \\ &= 0.67 \times 0.67 \times 0.5 \times 0.5 = 0.000112225 \end{aligned}$$

The process estimates the n-gram probability by dividing the observed frequency of a particular sequence by the observed frequency of a prefix. This ratio is called a relative frequency [59]. The above example provides an overview of how language models work. This task is automated by the SRILM (http://www.speech.sri.com/, accessed on 29 October 2021) tool.

We created seven language models using SRILM version 1.7.3, a toolkit for building and applying language models, primarily used in speech recognition, statistical tagging and segmentation, and machine translation. In previous work [60], it was possible to identify that the language models had a better performance than the bag of words (BoW) method, a technique widely used to represent text data. In [61], a study on short texts in different languages showed results of 82.4 precision on average. Recently, pre-trained models are using a deep learning approach [62]; however, their implementation requires sizeable computational processing capabilities. Language models have not been used for the task addressed in this work, as indicated by some studies cited in the related work.

The configuration used for the creation of the models was as follows:

- N-grams-2 (bigrams): the sequence of terms was determined after performing an analysis between N-grams of sizes 2, 4, and 6.
- We used Kneser–Ney discounting (kndiscount) as a smoothing method.
- Gtmin: specify the minimum counts for N-grams to be included in the language model, we used gt2min.

We used all the terms to train the seven models. Repeated text segments were also automatically removed to avoid overfitting. Then, the text was lemmatized with Freeling (http://nlp.lsi.upc.edu/freeling/, accessed on 30 October 2021) with the default config file for Spanish, which also allowed to obtain the part-of-speech (POS) tags for the text. Below we provide an example of input for the training process of the user story model "ver que la historia del usuario este enfocada a un asistente medico/see that the user's story is focused on a medical assistant":

ver VMIP1S0 que CS el DA0FP0 historia NCFP000 de SP usuario NCMS000 estar VMIP3P0 enfocar VMP00PF a SP uno DI0MS0 asistente NCCS000 médico AQ0MS00

see VB that IN the DT user NN 's story NN is VBZ focused VBN on IN a DT medical JJ assistant NN

Definition of grammatical class:
VB: Verb, base form/IN: Preposition or subordinating conjunction/DT: Determiner/NNs: Noun, plural/NN: Noun, singular or mass/VBZ: Verb, 3rd person singular present/VBN: Verb, past participle/JJ: Adjective

The example above shows the lemmatized words (word lemma) and their respective grammatical classes. The generated language models can be downloaded from https://github.com/gborrego/autotagsakagsd (accessed on 16 November 2021).

Finally, when the models were generated, each new text evaluated generates a numerical value that indicates the closeness of the text to the trained model. The following section details the use of language models.

4.3. Language Models and Slack Tagging Service Integration

Once we developed the language models, we integrated them with the Slack tagging service. To achieve this, we developed a REST web service in Python (recommendation service) to access the seven language models from any system capable of working with the HTTP protocol.

The operation of the seven language models together with the Slack tagging service is outlined in Figure 4, and described below:

1. An user clicks the button to start tagging a message, and the Slack tagging service does an HTTP request to the recommendation service sending the message text.
2. The recommendation service receives the text and sends it to each of the seven language models. Each model calculates a numerical value called perplexity, which expresses the confidence of the sentence tested in the language models. It is worth mentioning that the perplexity value is expressed between 0 and 1, where values close to zero indicate a powerful closeness. The measure used to compare the similarity in the language models was "average perplexity per word".
3. Once the recommendation service obtained the perplexity values of the seven language models; the values are ordered so that the first position represents the tag that fits better to the received text, according to the language models. Then, the service return to Slack an ordered list of tags, to be shown in the selection tag window (see Figure 2).

Although the described process seems cumbersome, the response time of the recommendation service is good enough not to cause any inconvenience to the end-user.

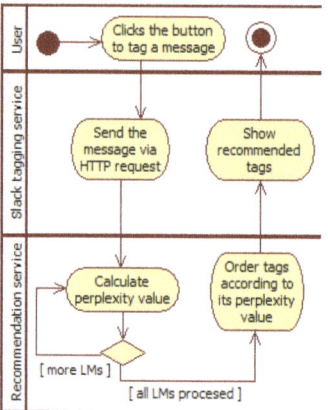

Figure 4. Activity diagram in unified modeling language, representing the integration of the Slack tagging service and the language models.

5. Materials and Methods

In this section, we present our method (based on the Wohlin et al. [63] guidelines) to evaluate the Slack tagging service. This method was based on the one followed in our past work [18]; even we reused part of the instrumentation in this experiment.

5.1. Objective

To analyze the use of the Slack tagging service, to determine the semantic correctness of the suggested tags concerning the context of the tagged message, as well as comparing the time spent to tag message in a both-way: assisted by auto-completion and assisted by NLP, from the point of view of professional developers and students in the context of agile and global software engineering.

5.2. Planning

According to the Wohlin et al. [63] guidelines, the planning phase consists of the following: experiment context definition, description of the subject selection, study design, variables and hypotheses definition, and a description of all the instruments required in the study.

5.2.1. Context and Subjects Selection

This study had two contexts: academic and industrial. The academic context was at a computer lab at two public universities. The industrial context was in different software development companies, selecting a room where they could not be interrupted during the evaluation activities.

We chose the subjects of both contexts by convenience. The participants in the academic context were students enrolled in either a software engineering program or an information technology program. We took care that they already had been in courses related to agile methodologies and web development since the evaluation's context scenario was a Web project driven by an agile method. The industrial context participants were professional software developers with experience in agile/distributed development and web application projects.

5.2.2. Study Design, Variables Selection, Furthermore, Hypotheses Formulation

This study is a quasi-experiment with a within-subject design because all the treatments were applied to all the participants. All the participants were mentally situated in a context scenario to interact through Slack with a counterpart; they used the tagging service (with predefined user tags) and followed a chatting guide that contains eight marks suggesting what to tag, without specifying the tag to use. There were two types of marks: one indicates using a tag recommended by the Slack tagging service, and another indicates selecting a tag from the tags catalog using the auto-complete feature.

The independent variable is represented by the different ways of choosing tags, using either the tags recommendation feature or the tags auto-completion feature. The dependent variables were the semantic correctness of the chosen tag (see Section 5.3.3 to know how we determined the tag correctness) and the time to select a tag. Besides, we observed the number of new tags that the participants created during the evaluation and the perceived ease of use and perceived usefulness. Based on the previous variables, we stated two hypotheses for this study.

- $H_{0Correctness}$. There is no significant difference in the number of correctly tagged messages by using the tags recommendation feature or the auto-completion feature.
- H_{0Time}. There is no significant difference in time spent tagging a message by using the tags recommendation feature or the auto-completion feature.

It is worthwhile to notice that the hypothesis $H_{0Correctness}$ corresponds to the RQ1 and the hypothesis H_{0Time} corresponds to the RQ2.

5.2.3. Instrumentation

Below, we present the instruments developed to conduct this study.

- Context scenario. This scenario was used in our past work [18], and it concerned two agile developers from different companies and locations working on the same project (medical appointments system), one of whom required information about a RESTful service that the other was developing. They had documentation debt, and consequently, they had to acquire the project AK by asking questions to each other. A complete description of this scenario is located at https://github.com/gborrego/autotagsakagsd (accessed on 16 November 2021).
- Chatting guides. Each pair of participants had to follow two guides (one per scenario role) to simulate a technical conversation using Slack regarding the context scenario. The guides had marks indicating when to tag using either the recommendation or auto-completion features. It is important to highlight that these guides were based on the chatting scripts used in our past work [18] and the interactions based on these scripts comprised the corpus with which we developed the language models of this study. Furthermore, it is worth remarking that using a chatting guide does not mean that we wrote exact phrases to be copied by the participants during their interactions. This chatting guide contained hints about what to request or respond to cause variations in the participants' writing, but with the same hints' semantic; in this way, we could test the robustness of the language models. Both chatting guides are located at https://github.com/gborrego/autotagsakagsd (accessed on 16 November 2021).
- Slack tagging service, and were presented in Section 3. The service also registers when it was activated and when it was closed to obtain the time spent tagging a message.
- Messages gathering program. It is a program developed in Node.js which uses the Slack API to extract the messages of public channels and then sends them to the Algolia repository.
- Extended TAM questionnaire. We prepared a questionnaire in Google Forms which was based on the Technology Acceptance Model [64] (TAM) using a Likert-7 scale. Just as in our previous work [18], we extended this questionnaire adding items such: how the Slack component integrates to the daily work, and another in which we asked about enhancements that the participants would consider adding to the component to answer the RQ3. Furthermore, this questionnaire collected the following demographic data: age, years of experience in agile development, years of experience in distributed/global software development. The complete questionnaire could be viewed at https://github.com/gborrego/autotagsakagsd (accessed on 16 November 2021).

5.3. Operation

In this part, we describe the four different stages of the study operation: the preparation stage, execution stage, data collection stage, and data validation stage.

5.3.1. Preparation

We verified that the computers to be used in this study had an Internet connection and a web browser that could execute the Slack messenger. We created a Slack workspace, and we installed the tagging service. Then, we added the user tags which correspond to the context scenario. The user tags were the same that we used in our previous study [18]: ServiceIP, RESTTest (related to TechnologicalSupport meta-tag); RestApikey, RestSecurity, Encryption, TestData, RESTResource, RESTResponse (related to Code meta-tag); AngularEncryption (related to Component meta-tag); and UserStory (related to Documentation). Furthermore, we checked that the tags corresponding to the trained NLP models appear in the respective list.

5.3.2. Execution

We had 52 participants which 30% were professional developers (26.4 years old on average, s.d. = 3.6) from three different Mexican companies, who declared an average of 2.7 years of experience in agile development (s.d. = 2.09) and 2.06 years (on average) of experience in global/distributed software development (s.d. = 2.5). The rest of the participants were students (21.6 years old on average, s.d. = 2.7) from two different Mexican universities. The evaluation sessions consisted of three parts which we explain below.

- Introduction (duration 10 min). We explained to the participants the study sessions along with their objectives. We organized the participants in pairs (to chat between them), and then we asked their email addresses to register them on the Slack workspace. We created a public channel for each pair of participants (to isolate the pairs' conversations), and we helped them complete the Slack registration. We gave the participants a short training session regarding how to use the tagging service (3 min, approx.), and they quickly explored the available tags (2 min, approx.). Then, we described the scenario in which they would be located to carry out the tasks, and we assigned a role to each pair member: either the developer working on a RESTful service or the developer who wished to use it. We gave them the corresponding guides, and we explained to them that the guides had marks indicating when to use the recommendation feature. Each pair member sat in a different part of the session room, ensuring they had no visual contact as if they were geographically distributed. We asked them to avoid talking to each other to emulate an environment of geographic distribution better.
- Interacting through Slack (duration 25 min). Following the corresponding guide, the participants used Slack to chat, and the tagging service aided them. We explained that they could paraphrase the messages since we gave them only a guide, not a script. We also told the participants that they could write a new tag (unregistered/invalid tag) if they could not find one that fitted a certain message on the options shown by the tagging service.
- Finalization (duration 3 min). When the participants had finished chatting through Slack, they answered the TAM-based questionnaire. Finally, we executed the message-gathering program to send all the channel conversations to a common repository.

5.3.3. Data Collection and Data Validation

The tagged messages from each conversation were collected from the repository, and we executed a script to determine whether the used tags were semantically correct or not. This process consisted of comparing the tagged messages against the expected tags, accounting that the meta-tags hierarchy of each tag was also considered correct. We noticed that participants also tagged messages that were not marked. Two expert persons manually evaluated whether the tag fitted the message semantics in those cases.

Besides, we extracted from the tagging service log file the closing and opening time of the tagging dialog to obtain the time spent to select a tag for each participant. Furthermore, we obtained the answers to the TAM questionnaire directly from Google Forms. Finally, we discarded those tagged messages that did not correspond to the chatting guide, such as the initial testing messages and those written after the final line of the chatting guide. The curated data set of the gathered messages is located in the following link: https://github.com/gborrego/autotagsakagsd (accessed on 16 November 2021).

6. Results

We obtained 356 tagged messages, where 50.6% were tagged using the auto-completion feature and 49.4% were tagged using the recommendation feature. We present the results organized as follows: one section per each stated hypothesis, another section in which we present the TAM results, and a section where we present additional observations of this study.

6.1. Tagging Correctness

In this part, it is worth remembering that the participants did not register any tag for this controlled study, as it is supposed to happen in a knowledge condensation implementation. For this reason, we allowed the participants to create tags when they considered that the predefined tags/meta-tags did not fit a message. We obtained that 14% of the messages were tagged with a tag created by the participants (new tags), from which 20% were incorrect, and 80% were correct. Furthermore, 25% of messages with user-created tags were considered synonyms of the predefined tags. Thus, to calculate the number of messages tagged correctly using the auto-completion feature, we summed the messages correctly tagged with the existing tags and those correctly tagged using a synonym tag. Moreover, we added the number of correctly tagged messages using the recommendation feature. It is worth remembering that we excluded the messages tagged with a new tag and included synonyms tags. Figure 5A shows that participants had more messages correctly tagged when they used the NLP recommendation feature.

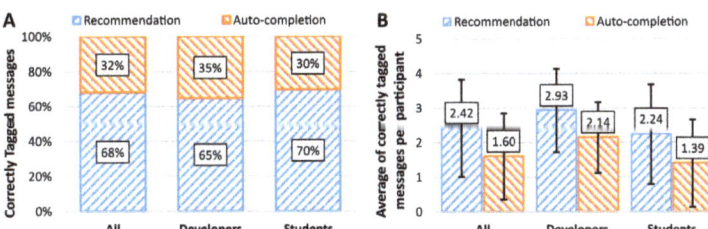

Figure 5. Tagging correctness using the recommendation feature and auto-completion feature, classified by participants profile. (**A**) Percentage of correct tagging on both ways. (**B**) Average of correct tagging by participant (whiskers represent standard deviation).

On the other hand, if we group the data by participants, the above-stated difference is also evident on the average of correctly tagged messages per participant on both participant profiles (see Figure 5B). We verified whether both data sets (average of correctly tagged messages using recommendation and auto-completion) were normally distributed by applying the Kolmogorov–Smirnov normality test. We obtained that the set of the tagged messages using the auto-completion feature did not fit the normal curve ($D = 0.18621$, p-value = 0.04731). Thus, we applied the Wilcoxon signed-rank test ($\alpha = 0.05$) to determine whether the difference between the average of correctly tagged messages per participant using recommendation and auto-completion feature was significant. We obtained that the difference is statistically significant ($W = 237.5$, p-value = 0.00452), hence, this result provides evidence to reject the null hypothesis $H_{0Correctness}$. With this result, we can answer the RQ1 saying that the participants tagged messages more coherently when they used the NLP recommendation.

6.2. Time Spent to Tag Messages

We obtained that the participants spent less time tagging when the recommendation feature was used ($avg = 11.03$ s, $std = 5.23$), either in general, or per each participant profile (see Figure 6A). In this case, we applied a T-test since both sets of data were normally distributed according to the Kolmogorov–Smirnov test (Time using recommendation: $D = 0.11737$, p-value = 0. 48686—Time using auto-completion: $D = 0.12835$, p-value = 0.34096). By applying T-test ($\alpha = 0.05$), we obtained that the difference between the time spent to tag by auto-completion and to tag using the recommendation feature was significant ($T = -5.10628$, p-value < 0.00001).

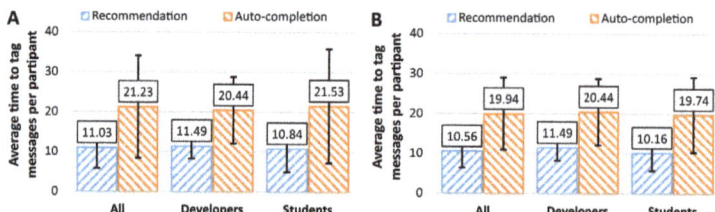

Figure 6. Time (in seconds) participants spent tagging a message using the recommendation and auto-completion features, classified by participants profile. (**A**) Results with outlier values, and (**B**) results without the outlier values (whiskers represent standard deviation).

To deeply explore this result, we discarded the outlier values of both data sets. Those values were obtained by applying the Grubbs' test with $\alpha = 0.05$ (Time using recommendation: $Z = 4.2314$, outlier-value = 33.18—Time using auto-completion: $Z = 5.0222$, outlier-value = 85.90); the difference was still visually present, as is shown in (Figure 6B). Removing the outlier values gives us that the average time was 10.56 s ($std = 4.13$) using the recommendation feature, and 19.94 s ($std = 9.06$) using auto-completion feature. We applied again T-test ($\alpha = 0.05$) and the difference was also statistically significant ($T = -6.48889$, p-value < 0.00001), thus, we can reject the null hypothesis H_{0Time}. With this result, we can answer the RQ2 saying that the participants tagged messages faster when they used the NLP recommendation.

6.3. Additional Observations

We noticed that 7.3% of the messages were tagged using the recommendation feature when it was not indicated in the guide, where 57% of the 7.3% were correctly tagged. It is important to remember that the language models were developed using tagged messages obtained from our previous study [18]. This fact indicates that the language models used to detect the context of the messages are robust enough to support different ways of writing mixed messages of the same topic. The average of correct and incorrect tagged messages per participant was 0.89 ($std = 0.90$) and 0.67 ($std = 0.68$), respectively. We applied the Kolmogorov–Smirnov test of normality and we obtained that both sets, correct and incorrect tagged messages, were normal (correct messages: $D = 0.23407$, p-value = 0.23742—incorrect messages: $D = 0.28579$, p-value = 0.0855). Hence, we applied a T-test ($\alpha = 0.05$), obtaining that the difference between both sets is not significant ($T = 0.83299$, p-value = 0.205331), which is an expected result since we did not train the language models to support the kind of messages where participants tagged.

In addition, when participants used the recommendation feature, and they tagged correctly, 85% of the time, the selected option was in the first position of the tag's list (see Figure 7A). This difference is significant according to the goodness of fit test based on χ^2 ($\alpha = 0.05$), assuming a distribution on which the first position has 75% and the rest

25% ($\chi^2 = 7.714$, p-value = 0.00548). On the other hand, when the participants used the recommendation feature, and they tagged incorrectly, 68% of the times the correct option was located in the first position of the list (see Figure 7B). This difference is significant according to the goodness of fit test based on $\chi^2(\alpha = 0.05)$, assuming a distribution on which the first position has 58% and the rest 42% ($\chi^2 = 4.599$, p-value = 0.032). This result means that the language models were preciser than the participants when tagging was unexpected. The correctly recommended tags were the following: DataTest, Documentation, TechnologicalSupport, AngularEncryption, RestResource, and UserStory.

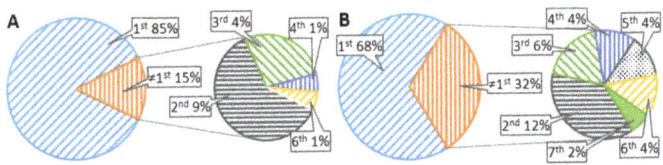

Figure 7. List position of the correct option on the recommendation feature when participants tagged correctly (**A**) and incorrectly (**B**) while using the recommendation component.

Finally, we obtained 176 messages tagged using the recommendation feature, without the unexpected tagged messages, wherein 112 messages the correct tag was in the first position of the recommendation list. With these data we can calculate a Precision = 112/176 = 0.6363 of NLP recommendation method. On the other hand, if we include the unexpected tagged messages, we have 204 tagged messages, wherein 125 messages the correct tag was in the first position of the recommendation list, i.e., now Precision = 125/204 = 0.6127. Furthermore, the difference between correct and incorrect recommendations is significant according to the goodness of fit test based on $\chi^2(\alpha = 0.05)$, assuming a distribution on which the correct recommendations has 56% and the rest 44% ($\chi^2 = 4.165$, p-value = 0.04126). The difference between correct and incorrect recommendations is also significant according to the goodness of fit test based on $\chi^2(\alpha = 0.05)$, assuming a distribution on which the correct recommendations has 54% and the rest 46% ($\chi^2 = 4.348$, p-value = 0.0371).

6.4. Qualitative Results

Participants responded to the TAM questionnaire after chatting through Slack using the chatting guide. We gathered all the questionnaire answers (based on Likert-7), and we applied the Cronbach alpha test to determine its consistency. We obtained $\alpha = 0.9641$ with 95% of confidence, which means that the TAM results present an excellent internal consistency. Analyzing the obtained results, participants perceived the implemented Slack tagging service as very useful and easy to use in general (median of 6 in Likert-7 scale) (see Table 2), highlighting that for both measures (usefulness and ease of use), the mode was 7 in Likert-7 scale. In Table 2, we observe that all the Likert values obtained for usefulness and ease of use represent a positive perception (greater of 4), which means that most of the participants perceived the tagging service as useful. However, there were some outlier values: developers' ease of use perception with values of 2, 3, and 4, and students' perception on each aspect, with 1. These low values could be related to the received suggestions about the pre-loaded tags; participants mentioned would have liked to participate in the tags registration. Furthermore, participants commented that the way of tagging must be improved; however, it entirely depends on the Slack conditions to create complements or services. Remarkably, participants suggested tagging before sending a message, not when the message has already been sent, which is how most tagging services currently work. Some comments of the participants accompany these results: "...*tagging messages is a bit tedious... I do not think that people would use it in an agile development...*", which indicates to us that the service operation has to be improved. Finally, we extended the TAM questionnaire with one question about how the tagging service integrates into daily work. We obtained that participants think that this service integrates easily to the working conditions of agile GSE projects. However, we obtained a value lower than 5 (Likert-7 scale) on quartile 25 and some outliers values (see Table 2), which could be related to the above-reported problems too. To summarize, we can state that participants perceived the Slack tagging service as very easy to use, useful and that it could be easily integrated into the daily work. With the previous statement, we answer the RQ3.

Table 2. Results of the extended TAM questionnaire, by participant profile. All the values are based on a Likert-7 scale.

	Students			Developers		
	Usefulness	Ease of Use	Integration	Usefulness	Ease of Use	Integration
Mode	7	7	7	7	7	7
Median	6	6	6	6	7	6
Q.75	7	7	7	7	7	7
Q.50	6	6	6	6	7	6
Q.25	5	5	5	5	6	4.25

7. Threats to Validity

To understand how valid the results are and how researchers and practitioners can use them, we present the validity threats according to Wohlin et al.'s [63] specification.

7.1. Conclusion Validity

This study had participants from different backgrounds and experiences using Slack. To balance the participants' knowledge about Slack, they received a brief training regarding using this application and how to use the tagging service. The participants were anonymous in this study. We did not know them before or after the evaluation; therefore, there was no reason to try to please us. Furthermore, we are aware that the evaluation period was relatively short; however, the results indicate a clear trend about the benefits of using the implemented Slack service.

7.2. Internal Validity

The study's sessions were short to prevent the participants from getting bored or tired. We attempted to avoid learning effects by using a counterbalancing technique, i.e., we placed the participants in groups and presented the conditions (auto-completion and recommendation) to each group in a different order. The TAM questionnaire results showed an excellent internal consistency when applying the Cronbach Alpha test. Besides, all the participants were volunteers and showed interest in collaborating in this study. Regarding persistence effects, the study was executed with subjects who had never taken part in a similar study. Moreover, the participants did not have any previous knowledge of the context scenario since it was fictitious.

7.3. Construct Validity

We measured the required time to tag a message using the tagging service, which registered when the tagging screen was shown and when it was closed by the participants. This time could have been affected by the participants' reading speed and by the time to decide a proper tag; however, we considered that participants had similar abilities, and the arrangement of the sets could have reduced this threat. To discover the participants' perceptions about how the tagging service integrates into their daily work, we extended the standard TAM questionnaire by adding one question with the same structure as TAM questions. This extension provided us with a structured means to obtain the participants' perceptions of topics that the conventional TAM questionnaire does not include.

7.4. External Validity

We identified two main threats to external validity: subjects and tasks/materials. Regarding subjects' threats, we included students to have more controlled conditions to conduct the study in an academic context. Unfortunately, these students had no experience with real agile GSE projects; however, they had already taken courses concerning agile software development and had been in touch with GSE topics during their university education. We included professional developers with experience in agile GSE to enforce external validity. Concerning threats about tasks and materials, the chatting guides were based on a fictitious scenario but included real-world situations and characteristics.

8. Discussion

This section discusses our results in two ways: implications about the AK condensation in agile GSE and natural language models. Finally, we present a summary of the contributions of this paper.

8.1. Implications about Architectural Knowledge Condensation in Agile Global Software Engineering

Literature reports that developers prefer to share AK with their counterparts using UTEM in agile GSE environments [10,19], mainly because of the language and culture differences. Thus, a mechanism to condense AK from UTEM logs is essential since these logs are the only source of AK many times [9,10]. A crucial part of the AK condensation is the classification mechanism, and it was implemented by using social tagging during UTEM interactions. Thus, efficient and coherent tagging is essential to retrieve AK in the future.

In this study, we observed that participants tagged coherently more frequently when they used the recommendation feature (RQ1). In our previous study [18], we obtained 62% of correctly tagged messages; in this study, we obtained 66%, where 32% corresponds to tagging by the auto-completion feature, and 68% corresponds to tagging by the recommendation feature. Thus, tag recommendations were essential to raising the correctness percentage. Regarding the spent time to tag a message, the use of the recommendation feature represents a significantly faster way to select tags (RQ2). In general, the time difference was nine second, which could represent the difference between a good or bad user experience [26], and consequently an essential factor to adopt the tagging service and the KC concept.

Regarding the perceived ease of use and the perceived usefulness, the Slack tagging service obtains high values of the Likert-7 scale, even on quartile 25, which means that participants consider the service very useful and usable (RQ3). Furthermore, participants considered that tagging service could be integrated into the daily work activities (RQ3). Further, there are points to enhance regarding tagging a message; however, it heavily depends on Slack's conditions to develop a complement. This situation could happen to any other UTEM since they have those conditions for developing plugins to complement its operation. We know that determining any software's usability implies more than applying questionnaires; usability testing includes observations, video-recording analysis, interviews, focus groups, etc. However, the main focus of this paper was to evaluate the Slack service in terms of time and coherence at tagging messages, and the TAM questionnaire was only to have a preliminary view of the Slack service's usability and usefulness.

The results showed that the semi-fixed tagging approach supported by the tags recommendation feature is an efficient (precise and fast) and well-perceived way to classify AK during UTEM interactions in an agile GSE environment. Furthermore, the results could reduce the cognitive load to agile GSE developers during their daily work (reducing the effort to think how to tag) and ease the AK finding since the AK is going to be aligned to the classification scheme, on which the tags are based. In this way, a developer could recover AK from a past project, using the tags and meta-tags to explore the project AK.

All the presented results led us to think that a recommendation feature could be essential for better AK classification when tagging is used as a classification mechanism (part of the AK condensation concept). This is because the tagging correctness percentage could increase in general; thus, the projects' AK could be better classified.

Considering the results of our study, we can state that the AK condensation concept could be implemented in the industry in the following way:

- We must implement a mechanism to determine how well a message was tagged. This implementation could run when developers look for projects' AK, using the searching mechanism proposed by the AK condensation concept; thus, when they find AK, they also could qualify the tagging. It is important to notice that the searching mechanism has as a source a repository where the UTEM logs will be stored; thus, it can contain messages from email, code repositories (commit comments), instant messengers, etc.
- Another mechanism could take from the messages repository the tagged ones with a good qualification; thus, determining when is a good qualification would be crucial.
- The same mechanism could group the messages by tag, and finally, pass the grouped messages to a final process to update the current language models or generate new ones based on new tags.

In this manner, a development team could have an updated set of language models, which would be helpful to suggest tags during UTEM interactions to condense AK.

8.2. Implications about Natural Language Models in Agile Global Software Engineering

The TAM evaluation provides encouraging results; developers and students identified the tag recommendation component as useful (based on language models and incorporated into the Slack tagging service) in software projects. The correctness percentage obtained by the recommender component gives evidence that the language models captured the sentence structure written by the participant. For instance, the language model identifies the word "story" with a probability of 0.75, and the phrase "user_story" is 0.76; that is, the terms used by the participants are very close to those used in the creation stage of the models.

However, for the terms that have a low probability, in some cases, they complement the probability of the grammatical class. For example, the model's question mark symbol ("?") has a probability of 0.05, but with its grammatical class, the probability increases to 0.97. The probabilities of the term and the grammatical class complement each other and give the tagging service robust performance, reflecting in the correctness percentage. The Kneser–Ney function softens those terms out of the vocabulary of the language model as a smoothing method. The precision results of the implemented models may indicate that using language models is promising. We observed that the small vocabulary of the corpus had a positive impact on the models, especially on tags where the frequency of repetition of the terms is high, such as "UserStory".

Concerning the precision of the language models, it could be considered a good score ($precision = 0.6363$ without including the unexpected tagged messages), taking into account the low amount of data (291 messages obtained from our previous study [18]) used to train the language models. For instance, works based on YouTube [47] or Stack-overflow [45,51,53] had available an extensive corpus (tagged posts and video tutorials of different years) to train the recommendation models. In particular, Parra et al. [47] reported that up to 66% of the tagged videos were considered relevant by developers. Thus, it could be interpreted that they obtained 0.66 of precision, which is pretty close to our precision value. Moreover, considering our acceptable results and our small corpus to train the language models, we could say that our approach presents certain tolerance to the "cold start" problem. However, we have to conduct more experiments to determine its robustness.

It is worth remarking that 68% of the times that participants tagged incorrectly, the correct option was the first element of the recommendation list. However, participants decided to choose another list element. This situation leads us to improve the language models of the recommendation feature to explore a mechanism of semi/auto-tagging. In this way, when a developer wants to tag a message, s/he could use this mechanism to assign a tag suggested by a language model.

We addressed the above-presented problem by training a model for each category due to the size of the corpus used to build them. However, each model provides the system with a refined and specific analysis. One aspect that we can identify to improve is incorporating more elements in the training corpus to reach a higher percentage of correct ones. Increasing the number of language models could leverage the operation of the Slack tagging service and the tagging process in an agile GSE environment. We created the models to support the Spanish language; however, we can build models for other languages by collecting a new corpus, for instance, English. Another rapid but less effective option would be to map our corpus to English, assuming that developers' communication is standardized. Finally, because of the success of our simple models, we did not investigate more complex recent models like BERT [62]. Such models can improve performance but would increase computational cost.

8.3. Summary of Contributions

The main contributions of our work could be summarized as four aspects, which are depicted below without a relevance order.

First, we implemented a Slack tagging service with an NLP recommendation-based feature that helps developers choose a correct tag regarding a conversation context in terms of AK. A coherent tagging is crucial to implementing the AK condensation concept. A coherent tagging would be the base to find AK when developers do not remember the exact phrases to search a UTEM interaction where AK was shared, accounting that UTEM logs usually are the only source of AK in agile GSE.

Second, the evaluation of the Slack tagging service provided statistical evidence to show that users tag faster ($T = -6.48889$, p-value < 0.00001) and with better semantic correctness, i.e., more coherently ($W = 237.5$, p-value = 0.00452), using the NLP recommendation feature than using the auto-completion one. As was mentioned above, coherent tagging is crucial to implement the AK condensation concept. Thus, having a service that helps accurately recommend relevant tags means a step towards industrial implementation of the AK condensation context. In the same sense, the implemented service allows to tag quickly, which could minimize the negative cost that interruptions cause [22,23], doing developers to redirect their attention from the main activity. Fast tagging represents a better usability perception, which also increases the software adoption [26].

Third, students and professional developers perceived the Slack tagging service as useful and easy to use. The evaluation was done using a TAM-based questionnaire, in which we obtained high Likert-7 values (e.g., 5 and 6) in low quartiles as Q.25. Furthermore, participants considered that the Slack tagging service could be easily integrated into the daily work of a software project. Although the participants' perception was good, we found aspects regarding the ease of use that could be improved if Slack offered more open elements for its extension. For the AK condensation concept, it is essential that the classification mechanism, implemented through the Slack tagging service, be embedded as naturally as possible in the developers' activities.

Fourth, the developed language models were robust enough to give coherent tag recommendations (participants tagged better using the NLP feature), even on parts of the chatting guide where we did not expect to be tagged. Thus, the developed language models could work with very different ways to write a given topic. This result encourages us to keep developing more language models to cover different AK topics to have diverse tag recommendations.

Fifth, we developed a corpus of 291 tagged messages contained in seven tags, which came from our past study [18]. This corpus could be used by academia to explore more possibilities to expand this work.

9. Conclusions

In agile GSE, a critical amount of AK is stored in the UTEM logs, and most of the time, these are the only AK source in a development team [9,10]. AK condensation is an approach that proposes a way to classify and recover AK from the UTEM logs. An implementation of this concept is based on tagging UTEM interactions to classify and ease AK retrieval.

In this paper, we presented the Slack tagging service (an implementation of an AK classification mechanism), which allows to tag messages based on a classification scheme obtained in a previous study [9]. The objective of this service was to help developers remember the available tags during their daily interactions, by auto-completing tags or by offering tags recommendations through NLP techniques and language models. Students and professional developers evaluated the Slack tagging service in a controlled experiment to compare both tagging mechanisms in terms of correctness and time to select a tag. The results statistically showed that more than 64% of the Slack messages were tagged correctly (i.e., tags were semantically related with the messages) using the tags recommendation, considering that the participants were unaware of the existing tags and the interaction context. Furthermore, when participants used the recommendation feature, they spent less time selecting a tag (statistically significant too), which is beneficial to the potential adoption of the Slack tagging service and the AK condensation concept. It is worth remarking that although the language models were based on a corpus in Spanish, we can use the same reported process to develop new language models based on a corpus in English, which is a common language in agile GSE interactions.

Further, the Slack tagging service was perceived as highly useful and usable, according to the TAM questionnaire results. However, we are aware that a more robust study is needed to determine the actual usability and user experience of the Slack service.

The obtained results are remarkable since the corpus volume to train the language models was significantly smaller than other approaches reported in the literature, which used sources as StackOverflow entries to train the recommendation models. Besides, our approach is focused on AK entirely dependent on software projects' circumstances and conditions, such as business rules clarifications, architectural decisions, and even deployment topics. Thus, we cannot use public Q&A services since these sources contain more general AK than the knowledge shared through UTEM during a software project.

Furthermore, our results reinforce the AK condensation concept feasibility, and that semi-fixed tagging could be an AK classification mechanism adequate for the agile GSE environment. Furthermore, these results lead us to conduct a long-term study in an agile GSE company to evaluate the concept adoption and evaluate how AK condensation could affect metrics such as time to finish tasks, interruptions quantity, team efficiency, and product quality. This long-term study will help us increase the training corpus's size for language models. Further, we will explore new ways to implement an AK classification mechanism, for instance, using ontologies or deep learning techniques. We will even explore new approaches to condense AK on any agile software development environment, either co-located or distributed or using new sources such as transcriptions, audio, or video, which are also used to share knowledge in agile GSE.

Another direction in our future work is to conduct a qualitative or mixed study to get the perception of software architects about the AK that could be condensed with our approach. This study should cover micro-architecture topics since our proposal focuses on the developers' daily interactions; however, this proposal could support architects' interactions, where macro-architecture topics could be more prevalent. Furthermore, developers could discuss architecture topics through UTEM that could be relevant at the macro-architecture level. Thus, a software architect's point of view will be relevant to determine the impact of the condensed AK, regardless its abstraction level.

Author Contributions: Conceptualization, Methodology, Software, Formal Analysis, Investigation, Writing—Original Draft, Writing—Review and Editing, Supervision, Project Administration, Visualization, G.B.; Conceptualization, Methodology, Software, Investigation, Data Curation, Writing—Original Draft, Writing—Review and Editing, S.G.-L.; Validation, Data Curation, Writing—Original Draft, Writing—Review and Editing, R.R.P. All authors have read and agreed to the published version of the manuscript.

Funding: This work was supported by the "Programa de Fortalecimiento a la Investigación 2021" (project numbers: *PROFAPI*_2021_0062 and *PROFAPI*_2021_0065) of the Instituto Tecnológico de Sonora. Furthermore, this work had support of the "Programa para el desarrollo profesional docente" for the first author (*ITSON-PTC*-114).

Institutional Review Board Statement: The study was approved by the Institutional Committee of Bioethics of the Instituto Tecnológico de Sonora (date of approval: 23 November 2020, approval code: 132).

Informed Consent Statement: Informed consent was obtained from all subjects involved in the study.

Data Availability Statement: All the data obtained in this study, as well as the generated languaje models, chatting scripts, and TAM questionnaire can be downloaded from https://github.com/gborrego/autotagsakagsd (accessed on 16 November 2021).

Acknowledgments: The authors are grateful to the participating companies: EMCOR, Sahuaro Labs and Tufesa, as well as the students of Instituto Tecnológico de Sonora and Universidad Tecnológica del Sur de Sonora, for the support provided to carry out the present study.

Conflicts of Interest: The authors declare no conflict of interest. The funders had no role in the design of the study; in the collection, analyses, or interpretation of data; in the writing of the manuscript, or in the decision to publish the results.

Abbreviations

The following abbreviations are used in this manuscript:

AK	Architectural Knowledge
GSE	Global Software Engineering
NLP	Natural Language Processing
TAM	Technology Acceptance Model
UTEM	Unstructured Textual and Electronic Media

References

1. Vliet, H.V. Software architecture knowledge management. In Proceedings of the IEEE 19th Australian Conference on Software Engineering (ASWEC 2008), Perth, Australia, 25–28 March 2008; pp. 24–31. [CrossRef]
2. Kruchten, P., Documentation of Software Architecture from a Knowledge Management Perspective—Design Representation. In *Software Architecture Knowledge Management: Theory and Practice*; Springer: Berlin/Heidelberg, Germany, 2009; Chapter 3, pp. 39–57. [CrossRef]
3. Dalkir, K. *Knowledge Management in Theory and Practice*, 2nd ed.; MIT Press: Cambridge, MA, USA, 2011; p. 485.
4. Holmstrom, H.; Conchuir, E.O.; Agerfalk, P.J.; Fitzgerald, B. Global Software Development Challenges: A Case Study on Temporal, Geographical and Socio-Cultural Distance. In Proceedings of the International Conference on Global Software Engineering (ICGSE '06), Florianopolis, Brazil, 16–19 October 2006; pp. 3–11. [CrossRef]
5. Zahedi, M.; Shahin, M.; Ali Babar, M. A systematic review of knowledge sharing challenges and practices in global software development. *Int. J. Inf. Manag.* **2016**, *36*, 995–1019. [CrossRef]
6. Beck, K.; Beedle, M.; van Bennekum, A.; Cockburn, A.; Cunningham, W.; Fowler, M.; Grenning, J.; Highsmith, J.; Hunt, A.; Jeffries, R.; et al. Manifesto for Agile Software Development. Web, 2001. Available online: https://agilemanifesto.org/ (accessed on 20 October 2021).
7. Cockburn, A.; Highsmith, J. Agile Software Development:The People Factor. *Computer* **2001**, *34*, 131–133. [CrossRef]
8. Yanzer Cabral, A.R.; Ribeiro, M.B.; Noll, R.P. Knowledge Management in Agile Software Projects: A Systematic Review. *J. Inf. Knowl. Manag. (JIKM)* **2014**, *13*, 1450010. [CrossRef]
9. Borrego, G.; Moran, A.A.L.; Palacio, R.R.; Rodriguez, O.M.O.; Morán, A.L.; Palacio, R.R.; Rodríguez, O.M.; Moran, A.A.L.; Palacio, R.R.; Rodriguez, O.M.O. Understanding architectural knowledge sharing in AGSD teams: An empirical study. In Proceedings of the 11th IEEE International Conference on Global Software Engineering, Irvine, CA, USA, 2–3 August 2016; pp. 109–118. [CrossRef]

10. Clerc, V.; Lago, P.; Vliet, H.V. Architectural Knowledge Management Practices in Agile Global Software Development. In Proceedings of the IEEE Sixth International Conference on Global Software Engineering Workshop, Helsinki, Finland, 15–18 August 2011; pp. 1–8. [CrossRef]
11. Tom, E.; Aurum, A.; Vidgen, R. An exploration of technical debt. *J. Syst. Softw.* **2013**, *86*, 1498–1516. [CrossRef]
12. Bider, I.; Söderberg, O. Becoming Agile in a Non-Disruptive Way. In Proceedings of the 18th International Conference on Enterprise Information Systems (ICEIS 2016), Rome, Italy, 25–28 April 2016; SCITEPRESS—Science and Technology Publications, Lda.: Setubal, Portugal, 2016; pp. 294–305. [CrossRef]
13. Bosch, J. *Software Architecture: The Next Step*; EWSA; Volume 3047, Lecture Notes in Computer Science; Oquendo, F., Warboys, B., Morrison, R., Eds.; Springer: Berlin/Heidelberg, Germany, 2004; pp. 194–199.
14. Holz, H.; Melnik, G.; Schaaf, M. Knowledge management for distributed agile processes: Models, techniques, and infrastructure. In Proceedings of the IEEE Enabling Technologies: Infrastructure for Collaborative Enterprises, Linz, Austria, 11 June 2003; pp. 291–294. [CrossRef]
15. Uikey, N.; Suman, U.; Ramani, A. A Documented Approach in Agile Software Development. *Int. J. Softw. Eng.* **2011**, *2*, 13–22.
16. Rios, N.; Mendes, L.; Cerdeiral, C.; Magalhães, A.P.F.; Perez, B.; Correal, D.; Astudillo, H.; Seaman, C.; Izurieta, C.; Santos, G.; et al. Hearing the Voice of Software Practitioners on Causes, Effects, and Practices to Deal with Documentation Debt. In *Requirements Engineering: Foundation for Software Quality*; Madhavji, N., Pasquale, L., Ferrari, A., Gnesi, S., Eds.; Springer International Publishing: Cham, Switzerland, 2020; pp. 55–70.
17. Ryan, S.; O'Connor, R.V. Acquiring and sharing tacit knowledge in software development teams: An empirical study. *Inf. Softw. Technol.* **2013**, *55*, 1614–1624. [CrossRef]
18. Borrego, G.; Morán, A.L.; Palacio, R.R.; Vizcaíno, A.; García, F.O. Towards a reduction in architectural knowledge vaporization during agile global software development. *Inf. Softw. Technol.* **2019**, *112*, 68–82. [CrossRef]
19. Estler, H..; Nordio, M.; Furia, C.A.; Meyer, B.; Schneider, J. Agile vs. Structured Distributed Software Development: A Case Study. In Proceedings of the 2012 IEEE Seventh International Conference on Global Software Engineering, Porto Alegre, Brazil, 27–30 August 2012; pp. 11–20.
20. Bagheri, E.; Ensan, F. Semantic tagging and linking of software engineering social content. *Autom. Softw. Eng.* **2016**, *23*, 147–190. [CrossRef]
21. Nagwani, N.K.; Singh, A.K.; Pandey, S. TAGme: A topical folksonomy based collaborative filtering for tag recommendation in community sites. In Proceedings of the 4th Multidisciplinary International Social Networks Conference (MISNC '17), Bangkok, Thailand, 17–19 July 2017. [CrossRef]
22. Bailey, B.P.; Konstan, J.A. On the need for attention-aware systems: Measuring effects of interruption on task performance, error rate, and affective state. *Comput. Hum. Behav.* **2006**, *22*, 685–708. [CrossRef]
23. Altmann, E.; Trafton, J.; Hambrick, Z. Momentary Interruptions Can Derail the Train of Thought. *J. Exp. Psychol. Gen.* **2014**, *143*, 215–226. [CrossRef]
24. Czerwinski, M.; Horvitz, E.; Wilhite, S. A Diary Study of Task Switching and Interruptions. In Proceedings of the SIGCHI Conference on Human Factors in Computing Systems (CHI '04), Vienna, Austria, 24–29 April 2004; Association for Computing Machinery: New York, NY, USA, 2004; pp. 175–182. [CrossRef]
25. Bailey, B.P.; Iqbal, S.T. Understanding Changes in Mental Workload during Execution of Goal-Directed Tasks and Its Application for Interruption Management. *ACM Trans. Comput.-Hum. Interact.* **2008**, *14*, 1–28. [CrossRef]
26. Nielsen, J. *Usability Engineering*; Interactive Technologies; Elsevier Science: Amsterdam, The Netherlands, 1994.
27. Kruchten, P.; Lago, P.; Vliet, H.V. Building Up and Reasoning About Architectural Knowledge. In Proceedings of the Second International Conference on Quality of Software Architectures, Västerås, Sweden, 27–29 June 2006; pp. 43–58._8. [CrossRef]
28. Bass, L.; Clements, P.; Kazman, R. *Software Architecture in Practice*, 2nd ed.; Addison-Wesley Professional: Boston, MA, USA. 2003; p. 560.
29. ISO/IEC/IEEE. Systems and Software Engineering—Architecture Description. In *ISO/IEC/IEEE 42010:2011(E)*; IEEE: New York, NY, USA, 2011. [CrossRef]
30. Kruchten, P. *The Rational Unified Process: An Introduction*, 3rd ed.; Addison-Wesley Longman: Reading, MA, USA, 2003.
31. Clements, P.; Kazman, R.; Klein, M. *Evaluating Software Architectures: Methods and Case Studies*; SEI Series in Software Engineering; Addison-Wesley: Boston, MA, USA, 2001.
32. Vogel, O.; Arnold, I.; Chughtai, A.; Kehrer, T. *Software Architecture: A Comprehensive Framework and Guide for Practitioners*; Springer: New York, NY, USA, 2011. [CrossRef]
33. Dorairaj, S.; Noble, J.; Malik, P. Knowledge Management in Distributed Agile Software Development. In Proceedings of the IEEE 2012 Agile Conference, Dallas, TX, USA, 13–17 August 2012; pp. 64–73. [CrossRef]
34. Jiménez, M.; Piattini, M.; Vizcaíno, A. Challenges and Improvements in Distributed Software Development: A Systematic Review. *Adv. Softw. Eng.* **2009**, *2009*, 710971. [CrossRef]
35. Awar, K.; Sameem, M.; Hafeez, Y. A model for applying Agile practices in Distributed environment: A case of local software industry. In Proceedings of the 2017 International Conference on Communication, Computing and Digital Systems (C-CODE 2017), Islamabad, Pakistan, 8–9 March 2017; pp. 228–232. [CrossRef]
36. Sneed, H.M. Dealing with Technical Debt in agile development projects. In Proceeding of the 6th International Conference (SWQD 2014), Vienna, Austria, 14–16 January 2014; pp. 48–62. [CrossRef]

37. Clear, T. Documentation and Agile Methods: Striking a Balance. *SIGCSE Bull.* **2003**, *35*, 12–13. [CrossRef]
38. Souza, E.; Moreira, A.; Goulão, M. Deriving architectural models from requirements specifications: A systematic mapping study. *Inf. Softw. Technol.* **2019**, *109*, 26–39. [CrossRef]
39. Mirakhorli, M.; Chen, H.M.; Kazman, R. Mining Big Data for Detecting, Extracting and Recommending Architectural Design Concepts. In Proceedings of the 1st International Workshop on Big Data Software Engineering, Florence, Italy, 23 May 2015; pp. 15–18. [CrossRef]
40. Shahbazian, A.; Kyu Lee, Y.; Le, D.; Brun, Y.; Medvidovic, N. Recovering Architectural Design Decisions. In Proceedings of the 15th International Conference on Software Architecture, Seattle, WA, USA, 30 April–4 May 2018; pp. 95–104. [CrossRef]
41. Sobernig, S.; Zdun, U. Distilling architectural design decisions and their relationships using frequent item-sets. In Proceedings of the 13th Working IEEE/IFIP Conference on Software Architecture, Venice, Italy, 5–8 April 2016; pp. 61–70. [CrossRef]
42. Treude, C.; Storey, M.A. Work item tagging: Communicating concerns in collaborative software development. *IEEE Trans. Softw. Eng.* **2012**, *38*, 19–34. [CrossRef]
43. Storey, M.A.; Ryall, J.; Singer, J.; Myers, D.; Cheng, L.T.; Muller, M. How software developers use tagging to support reminding and refinding. *IEEE Trans. Softw. Eng.* **2009**, *35*, 470–483. [CrossRef]
44. Al-kofahi, J.M.; Tamrawi, A.; Nguyen, T.T.; Nguyen, H.A.; Nguyen, T.N. Fuzzy Set Approach for Automatic Tagging in Evolving Software. In Proceedings of the 2010 IEEE International Conference on Software Maintenance (ICSM), Timisoara, Romania, Timisoara, Romania, 12–18 September 2010; pp. 1–10. [CrossRef]
45. Liu, J.; Zhou, P.; Yang, Z.; Liu, X.; Grundy, J. FastTagRec: Fast tag recommendation for software information sites. *Autom. Softw. Eng.* **2018**, *25*, 675–701. [CrossRef]
46. Mushtaq, H.; Malik, B.H.; Shah, S.A.; Siddique, U.B.; Shahzad, M.; Siddique, I. Implicit and explicit knowledge mining of Crowdsourced communities: Architectural and technology verdicts. *Int. J. Adv. Comput. Sci. Appl.* **2018**, *9*, 105–111. [CrossRef]
47. Parra, E.; Escobar-Avila, J.; Haiduc, S. Automatic tag recommendation for software development video tutorials. In Proceedings of the 2018 IEEE/ACM 26th International Conference on Program Comprehension (ICPC), Gothenburg, Sweden, 27 May–3 June 2018; pp. 222–232. [CrossRef]
48. Jovanovic, J.; Bagheri, E.; Cuzzola, J.; Gasevic, D.; Jeremic, Z.; Bashash, R. Automated semantic tagging of textual content. *IT Prof.* **2014**, *16*, 38–46. [CrossRef]
49. Wang, T.; Wang, H.; Yin, G.; Ling, C.X.; Li, X.; Zou, P. Tag recommendation for open source software. *Front. Comput. Sci.* **2014**, *8*, 69–82. [CrossRef]
50. Alqahtani, S.S.; Rilling, J. An Ontology-Based Approach to Automate Tagging of Software Artifacts. In Proceedings of the International Symposium on Empirical Software Engineering and Measurement, Toronto, ON, Canada, 9–10 November 2017; pp. 169–174. [CrossRef]
51. Zhou, P.; Liu, J.; Liu, X.; Yang, Z.; Grundy, J. Is deep learning better than traditional approaches in tag recommendation for software information sites? *Inf. Softw. Technol.* **2019**, *109*, 1–13. [CrossRef]
52. Martins, E.F.; Belém, F.M.; Almeida, J.M.; Gonçalves, M.A. On cold start for associative tag recommendation. *J. Assoc. Inf. Sci. Technol.* **2016**, *67*, 83–105. [CrossRef]
53. Soliman, M.; Galster, M.; Salama, A.R.; Riebisch, M. Architectural knowledge for technology decisions in developer communities: An exploratory study with StackOverflow. In Proceedings of the 13th Working IEEE/IFIP Conference on Software Architecture, Venice, Italy, 5–8 April 2016; pp. 128–133. [CrossRef]
54. Musil, J.; Ekaputra, F.J.; Sabou, M.; Ionescu, T.; Schall, D.; Musil, A.; Biffl, S. Continuous Architectural Knowledge Integration: Making Heterogeneous Architectural Knowledge Available in Large-Scale Organizations. In Proceedings of the IEEE International Conference on Software Architecture, Gothenburg, Sweden, 3–7 April 2017; pp. 189–192. [CrossRef]
55. Lin, B.; Zagalsky, A.E.; Storey, M.A.; Serebrenik, A. Why Developers Are Slacking Off: Understanding How Software Teams Use Slack. In Proceedings of the 19th ACM Conference on Computer Supported Cooperative Work and Social Computing Companion, San Francisco, CA, USA, 27 February–2 March 2016; pp. 333–336. [CrossRef]
56. White, K.; Grierson, H.; Wodehouse, A. Using Slack for Asynchronous Communication in a Global Design Project. In Proceedings of the International Conference on Engineering and Product Design Education, Oslo, Norway, 7–8 September 2017; pp. 346–351.
57. Chatterjee, P.; Damevski, K.; Pollock, L. Exploratory Study of Slack Q&A Chats as a Mining Source for Software Engineering Tools. In Proceedings of the 2019 IEEE/ACM 16th International Conference on Mining Software Repositories (MSR) (ICSE 2019), Montreal, QC, Canada, 26–27 May 2019; MSR 2019 MSR Technical Papers; pp. 490–501.
58. Martínez-García, J.R.; Castillo-Barrera, F.E.; Palacio, R.R.; Borrego, G.; Cuevas-Tello, J.C. Ontology for knowledge condensation to support expertise location in the code phase during software development process. *IET Softw.* **2020**, *14*, 234–241. [CrossRef]
59. Jurafsky, D.; Martin, J.H. *Speech and Language Processing*, 2nd ed.; Prentice-Hall: Upper Saddle River, NJ, USA, 2009.
60. González-López, S.; Borrego, G.; López-López, A.; Morán, A.L. Clasificando conocimiento arquitectónico a través de técnicas de minería de texto. *Komput. Sapiens* **2018**, *1*, 29–33.
61. Vatanen, T.; Väyrynen, J.J.; Virpioja, S. Language Identification of Short Text Segments with N-gram Models. In Proceedings of the Seventh International Conference on Language Resources and Evaluation (LREC'10), Marseille, France, 11–16 May 2010; European Language Resources Association (ELRA): Valletta, Malta, 2010.
62. Devlin, J.; Chang, M.; Lee, K.; Toutanova, K. BERT: Pre-training of Deep Bidirectional Transformers for Language Understanding. *arXiv* **2018**, arXiv:1810.04805.

63. Wohlin, C.; Runeson, P.; Hst, M.; Ohlsson, M.C.; Regnell, B.; Wessln, A. *Experimentation in Software Engineering*; Springer: Berlin/Heidelberg, Germany, 2012.
64. Davis, F.D. Perceived Usefulness, Perceived Ease of Use, and User Acceptance of Information Technology. *MIS Q.* **1989**, *13*, 319–340. [CrossRef]

Article

The Next Generation of Edutainment Applications for Young Children—A Proposal

Adriana-Mihaela Guran *,†, Grigoreta-Sofia Cojocar † and Laura-Silvia Dioşan †

Department of Computer Science, Babeş-Bolyai University, 400085 Cluj-Napoca, Romania; grigo@cs.ubbcluj.ro (G.-S.C.); lauras@cs.ubbcluj.ro (L.-S.D.)
* Correspondence: adriana@cs.ubbcluj.ro
† These authors contributed equally to this work.

Abstract: Edutainment applications are a type of software that is designed to be entertaining while also being educational. In the current COVID-19 pandemic context, when children have to stay home due to the social distancing rules, edutainment applications for young children are more and more used each day. However, are these applications ready to take the place of an in-person teacher? In this paper, we propose a new generation of edutainment applications that are more suitable for preschoolers (aged 3–6 years old in our country) and closer to the in-person student–teacher interaction: emotions aware edutainment applications. We discuss the most important challenges that must be overcome in developing this kind of applications (i.e., recognizing children's emotions, enhancing the edutainment application with emotion awareness, and adapting the interaction flow) and the first steps that we have taken for developing them.

Keywords: edutainment applications; emotions aware applications; adaptive learning; software architecture; services

Citation: Guran, A.-M.; Cojocar, G.-S.; Dioşan, L.-S. The Next Generation of Edutainment Applications for Young Children—A Proposal. *Mathematics* **2022**, *10*, 645. https://doi.org/10.3390/math10040645

Academic Editors: Radu Tudor Ionescu and Florin Leon

Received: 23 December 2021
Accepted: 16 February 2022
Published: 19 February 2022

Publisher's Note: MDPI stays neutral with regard to jurisdictional claims in published maps and institutional affiliations.

Copyright: © 2022 by the authors. Licensee MDPI, Basel, Switzerland. This article is an open access article distributed under the terms and conditions of the Creative Commons Attribution (CC BY) license (https://creativecommons.org/licenses/by/4.0/).

1. Introduction

The term edutainment is a mixture between *education* and *entertainment* [1–3]. Edutainment applications are a type of software that is designed to be entertaining while also being educational. Various studies have analyzed the impact that the use of edutainment applications has on the learning outcome [4,5]. The studies' results show that using edutainment applications in the classrooms influences positively the learning outcome. Edutainment applications for very young children (aged between 3–6 years old) need to integrate the learning goals with young children's main activity, play. These applications are meant to help educators in teaching and consolidating new knowledge, especially now, when society is facing multiple challenges due to the crisis introduced by the COVID-19 pandemic, and most educational activities are performed using digital approaches (online or offline). Modern educational approaches include a broad range of technology-enhanced educational strategies to provide support for digital learning. Existing digital technology can assist students in learning, and it can play a crucial role in the field of education, but it does not have enough capabilities to replace the teachers. A teacher is not just a facilitator of knowledge, but also a guide, a mentor and an inspiration for students. Teachers do more than just the one-way task of instructing students. They can recognize social cues that would be impossible for a machine to identify, especially non-verbal or invisible (natural) interactions, that affect the learning experience. These cues help recognize students' difficulties that might be more personal or emotional in nature, and which a machine cannot identify. The teachers also help to contextualize lessons in real-time, which might not be possible for a piece of technology to do. The human interaction cannot be replaced by computers, and human skills like decision-making or time management cannot be taught by technology. Technology by no means can be a replacement for teachers, but it can be used effectively to enhance the learning process.

Because in the current context of society, technology not only supports the learning process but sometimes even replaces the in-person student–teacher interaction, we should try to empower technology with some of the skills the teachers naturally have. Recognizing emotions during the learning process is an ability that teachers have, but one that the interactive applications usually do not possess, even though emotions have a high impact on the results of the learning process [6–8].

The current trends in education are to include technology. Moreover, due to the current pandemic context, the educational system has been forced to replace physical interaction with remote learning with the support of technology. While for school children solutions to continue the learning process have been found, for preschoolers the lack of digital resources and the lack of digital competencies have brought difficulties. Developing digital applications for preschoolers should provide the learning content, but it should also be able to mediate the interaction of children with technology, by being aware, at least, of the children's emotional state. In this paper, we propose a new type of edutainment applications for young children, that takes into consideration the emotions of the young user during the interaction and adapts when a negative emotion is identified.

The idea of enhancing learning supporting tools with emotion awareness and adapting their interaction flow based on the emotions of the learner is not new and there are various studies available in the field. Feidakis provides a summary of emotion-aware systems designed for e-learning in virtual settings in [9]. Another study by Ruiz et al. [10] suggested a method for assessing students' mood based on a model of twelve positive and negative emotions, using self-report and observing interactions with teachers. However, the existing studies and approaches have been validated through case studies on university students and in the context of e-learning systems, and most of them do not automatically identify the learner's emotions while interacting with the application. Usually, the user's emotions are identified based on some questionnaires or other kind of input from the user. Our proposal is different because it is addressed toward young children, aged 3–6 years old, that have fast changes of emotional state. The very young age of these users creates difficulties in automatic emotion recognition. Furthermore, other existing methods used for emotion recognition, like self-reporting, cannot be applied to this type of users.

The main contribution of this paper is to propose a new generation of edutainment applications for young children (aged 3–6 years old): emotions aware edutainment applications. We present the need for such applications, the advantages that they bring, the challenges that must be overcome in order to develop such applications, a possible architecture, a three-phase process for developing them, and a very simple prototype.

The rest of the paper is structured as follows. Section 2 presents the concept of edutainment applications and the challenges in designing them for young children. In Section 3, we present our proposal for the next generation of edutainment applications: emotions aware edutainment applications. We describe the existing challenges and possible solutions. A proof of concept prototype is presented in Section 4. Discussions are given in Section 5. The paper ends with further work (Section 6).

2. Edutainment Applications for Young Children

Educational entertainment, or *edutainment*, has been used to present teaching content in an entertaining context. Edutainment blends games with learning and provides a fun and enjoyable way of acquiring new knowledge. However, designing edutainment applications for young children involves many challenges like wrapping the educational content in games, deciding the appropriate interaction for young children, providing a balance between learning and fun, handling errors in interaction, or addressing failures when an incorrect answer is provided.

In general, designing applications for young children is a challenging task, and the existing guidelines refer to children having between 0 and 8 years as only one kind of user [11,12]. However, there are significant differences among children in various age groups. Children aged 3 to 6 years old cannot read or write, so interaction using written messages is not recommended, they need adult guidance and monitoring, and to keep them focused, they need rewards for their actions. Designing edutainment applications for such young children imposes the design of a game-based learning strategy with new constraints on interaction (for example, no written input/output), appropriate feedback based on child's performance, and appropriate rewards based on the obtained results.

Presently, an edutainment application for young children consists of a predefined set of tasks $\{T_1, T_2, ..., T_n\}$, that a child must execute in order to gain new knowledge or new skills. Each task has a *difficulty level* and a *type*. Usually the difficulty level is easy, medium or complex, and the type of the task is a quiz, memory-game, riddle or puzzle. Currently, the edutainment applications are following a sequential flow (that we will call the *normal flow*), where the content and the tasks are presented either in a predefined order (meaning that if we run the application multiple times, the tasks will always be presented in the same order) or in a random order (meaning that if we run the application multiple times, some tasks will be presented in a different order). If a child cannot perform task T_i, then the application proposes task T_{i+1}, abandoning T_i. Some applications will replay the tasks that have been skipped at the end of the predefined set of tasks (after T_n). As these edutainment applications are meant for young children, aged between 3–6 years old, the entire execution time of such an application does not exceed 10 min.

For these applications, challenges occur when a child does not perform the task correctly or when the child decides to quit interacting with the edutainment. In the situation of wrong answers, the edutainment application should provide hints to help the child perform the task. New questions arise in this situation based on the time to wait for the child's answer, when to provide hints, how to provide the hints, how many times hints should be presented before deciding to move on with the interaction. In the classical (in-person) teaching-learning scenario, these challenges are gently solved by the educational experts by providing encouraging feedback or hints for task accomplishment, or by proposing a different task that can be successfully accomplished by the child. However, compensating for the teacher's absence when difficulties occur is almost impossible if there is no additional information about a child's actions or reactions.

3. Next Generation of Edutainment Applications for Young Children

The current generation of edutainment applications for young children are focusing on providing good entertaining and education aspects, but the next generation of edutainment applications should also consider a child's emotional state during interaction. It is important to start developing emotions aware edutainment applications for young children as the children' emotional state influences the learning outcome. Furthermore, emotion-aware edutainment application can be used to support learning at a child's own pace.

In our vision, the next generation of edutainment applications should allow a customized approach for task selection when difficulty in performing a task is encountered or when the child's emotional state changes to a negative emotion, such as frustration or anger. However, enhancing edutainment applications with emotions awareness is not an easy task and various challenging aspects must be tackled. Presently, the most important challenges are as follows:

- How to automatically identify a child's emotions?
- How to integrate emotions recognition within an edutainment application?
- How to adapt the edutainment application's interaction flow based on the identified emotions?

In the following subsections we address each of these challenges. First, we describe what emotions are, how they can affect the learning process, and how they can be automatically detected. Then, we give a description of the architecture and development process that we propose for enhancing edutainment applications with emotions recognition. Afterwards we describe the algorithms that we propose for adapting the interaction based on the identified emotions.

3.1. Young Children's Emotions

In the literature, various definitions of emotions have been given. Any brief episodes of coordinated changes (brain, autonomic, and behavioral) that facilitate a reaction to an important event are classified as emotions. Emotions are also targeted, according to Frijda, and require a relationship between the individual experiencing the feeling and the emotion's object [13,14]. Davou defines emotions as the organism's reaction to any disturbance of the perceptual environment [15]. Among all emotions, there are some *basic emotions* which are patterns of physiological reactions and which can be easily recognized universally. A few examples of basic emotions are fear, anger and happiness [16–19].

Understanding emotions, managing emotions and empathizing with others are all important skills in nonverbal communication and social integration. Emotions are also components of school readiness and academic success [6,7]. Researchers have found that there are statistically significant associations between social-emotional skills measured in kindergarten and key young adult outcomes over multiple domains of education, mental health, employment, substance use and criminal activity. Non-cognitive skills interact with cognitive skills to enable success in school and at the workplace. Pekrun [20] identified the so-called *academic emotions* and discovered that a good mood encourages comprehensive, creative thinking (Figure 1). Negative emotions like anger, sadness, fear or boredom are negatively associated to the learning process and outcomes, whereas positive emotions like enjoyment and hope are positively related to the learning process and outcomes. In many cases, negative emotions are also detrimental to motivation, performance and learning [21–23].

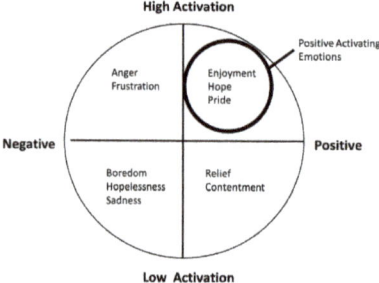

Figure 1. Academic emotions [20].

Children communicate their emotions through multiple channels like gestures, vocalization, body posture, body movements and facial expressions [24]. Frustration is a common emotion in young children that occurs when children cannot achieve a specified goal. Frustration is a healthy and normal feeling that can help a youngster learn more effectively. Frustration indicates that the child should find another solution to the problem encountered. Still, frustration must be handled before changing to anger or tantrum. For the next generation of edutainment applications for young children we are interested in identifying negative emotions like fear, anger or boredom that appear during interaction and that could negatively impact the learning process. Other (positive) emotions like happiness or surprise are also important, as they can be used to assess the satisfaction of the child while interacting with the application.

3.2. Automatic Child Emotion Recognition

Children and adults express their feelings through facial expressions, through their body, their behavior, their words and gestures. Due to the varying ways of expressing emotions, it is not easy to automatically identify a person's emotions. Most research has focused on automatic identification of emotions from facial expressions, which involves two steps: face detection and emotion recognition. Because automatic face localization is a required stage of facial image processing for many applications, face detection research is very advanced [25,26], having a 99% accuracy [27], but the emotion recognition (or identification) from facial expressions still remains an open problem. The Machine Learning (ML) approaches achieved only around 75% accuracy for facial emotion classification [28,29].

Datasets with children's faces are challenging to create since image gathering requires parental authorization, and discovering and documenting pupils takes time and effort. There are only a few datasets available in this context, and they are not evenly distributed in terms of emotions. The most used datasets for emotion recognition from faces are CAFE [30], CK++ [31], FER [32] and JAFFE [33]. However, only CAFE and FER contain children images, while the other datasets contain only adult images. Even in these datasets, there are some aspects that negatively impact the accuracy of the results, like the small number of children images, the lack of natural expressions in images as there are only posed images in the datasets, and the imbalanced representation of some emotions. In these datasets, the predominant emotions are neutral, sad and happy.

Automatic identification of children's emotions from facial expressions is even more challenging due to additional factors, like lack of available datasets with children faces and the ways children react when they know pictures are taken with them. In [34], we have conducted a pilot study in order to determine the appropriateness of several ML-algorithms for children's emotions automatic identification from their facial expressions, using different datasets (composed of adults and children faces). In different projects, several teams of 3–5 students from our faculty have implemented various methods for emotion recognition in images and videos. Five different projects have used Convolutional Neural Networks (CNN) [35], but with different network architectures. In order to determine the best hyper-parameters of the emotion classifier, each CNN architecture was trained in a cross-validation framework (by a random division of training data into learning and validation parts). Four projects used the same dataset, FER, and the same type of photos to train and test the classifier (images with adults). For the training–testing flow, different scenarios were used: some teams trained the classifier on images of adults and tested it on the same type of images or on a mixed dataset (adults and kids), one team trained and tested the classifier on images of kids, and others trained and tested the classifier on mixed images (adults and kids). The obtained accuracy is different for each project, ranging from 44% to 81%. The different performances obtained by the projects could have been caused by the distinct training setups. The obtained results show that an emotion classifier trained on adults images can be successfully used to other adult images (obtaining the best accuracy of 81%), but the classifier's accuracy decreases when used on mixed images (adults and children) to 50%. The project that has trained and tested the CNN with only children images taken from the CAFE dataset, has obtained an accuracy of 68%.

3.3. Integrating Emotions Recognition into an Edutainment Application

In order to enhance edutainment applications with emotions recognition capabilities we have to add new modules to them. In our opinion, the enhanced edutainment application should contain at least the following modules:

- the *edutainment* module—that is still responsible for presenting the learning content and the tasks to aid the comprehension of the new knowledge;
- an *emotion recognition* module—that is responsible for the identification of the emotional state of the user;
- a *coordinator* module—that is responsible for the coordination of the other modules;
- a *dataset building* module—that is responsible for the datasets creation and management (for example adding images of young children and annotating them with the corresponding emotion).

The edutainment and emotion recognition modules should run in parallel and they should communicate via the coordinator. There are at least two possible scenarios for when information should be exchanged between these modules:

1. The edutainment module, at some predefined moments from the tasks' execution flow, sends requests about the emotional state of the user (e.g., when a task is finished,

when a long time has passed since using the application, when the user has difficulties in completing a task, etc.) and it adapts the interaction based on the received response.
2. The emotion recognition module continuously sends information about the identified emotions to the edutainment module. Based on the received information, the edutainment module will filter the negative emotions and their context, and will trigger its interaction adaptation.

In Figure 2, a high level view of an edutainment application enhanced with emotion recognition capabilities considering the second scenario is shown.

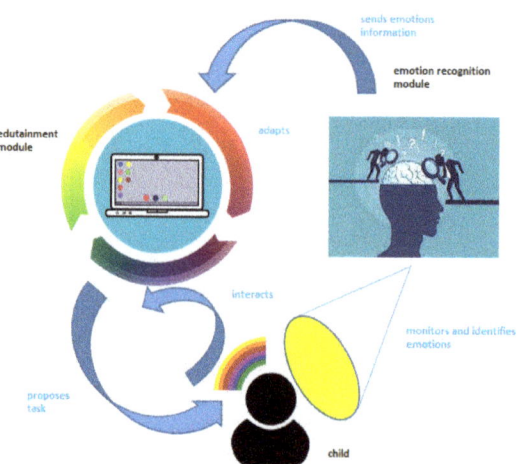

Figure 2. High-level view of an edutainment application enhanced with emotion recognition.

Both scenarios have advantages and disadvantages. The first scenario is more efficient as the information exchange takes places only at predefined moments, but it may ignore important emotional changes that can occur between two successive moments when the emotions are being recognized by the corresponding module. The second scenario is more resource expensive as the emotion recognition module needs to continuously identify emotions and to send them to the edutainment module. In this scenario, the edutainment module has to both play the exposed content and, at the same time, check the received information. In this regard, it is important to mention that the emotional state of a child may change very fast, that is why the information sent from the emotion recognition module to the edutainment module may lead to bottlenecks.

For this proposal we have decided to use a variation of the first scenario, as very often these applications will be used on computers with limited resources. We propose to use the following phases for developing emotions recognition enhanced edutainment applications:

- **Phase 1**—*Development of the edutainment module.* In this step, together with the other stakeholders (educational experts, kindergarten teachers, etc.) should be decided the tasks to be included in the edutainment module, their difficulty level, their type, and the normal interaction flow. The idea of each application may be decided by the early childhood educators. Each application should address one or multiple domains from the curricula and should be composed by a learning part, where new content is presented, and a practical part that contains tasks to support knowledge fixation. One possible approach that can be used for the edutainment module development is described in [36].
- **Phase 2**—*Development of the emotion recognition module and dataset building.* After the edutainment module was developed, a child's emotion recognition approach must be selected and validated. If the accuracy of the selected approach is not the desired one, new datasets with data about children while interacting with the edutainment

module could be created in order to improve the accuracy of the emotion recognition approach. These datasets should be annotated by human experts. If there already exist big enough datasets, or if the accuracy of the emotion recognition approach used is good enough for this type of applications, then the dataset building step may be skipped.
- **Phase 3**—*Integration of emotions awareness into the edutainment module.* The edutainment module should be modified in order to also include the adaptation feature for the interaction flow.

In this approach each of the proposed modules should provide at least the following services (presented in Figure 3):

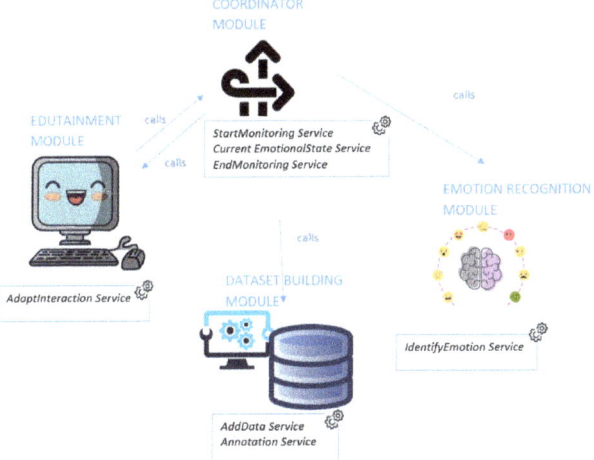

Figure 3. Proposed modules and services.

- The *IdentifyEmotion Service*, part of the emotion recognition module, that is responsible for identifying the emotional state from the data sent to it (images, etc.).
- The *AddData Service*, part of the dataset building module that is responsible for adding new data to the datasets used for training and validation of the selected emotion recognition approach.
- The *Annotation Service*, part of the dataset building module that is used to annotate the data from the datasets.
- The *StartMonitoring Service*, part of the coordinator module that is responsible for initiating the monitoring activity of the emotional state of the child.
- The *EndMonitoring Service*, part of the coordinator module that is responsible for ending the monitoring activity, started by the *StartMonitoring Service*.
- The *CurrentEmotionalState Service*, part of the coordinator module that is responsible for obtaining the necessary input data for the emotion recognition approach and sending it to the *IdentifyEmotion Service* in order to obtain the current emotional state of the child.
- The *AdaptInteraction Service*, part of the edutainment module that is responsible for initiating adaptation of the interaction flow, in order to change the child's emotional state.

In the following we describe how these services should collaborate in order to add emotion awareness to the edutainment module. The UML activity diagram from Figure 4 shows how the proposed services collaborate, after the application starts.

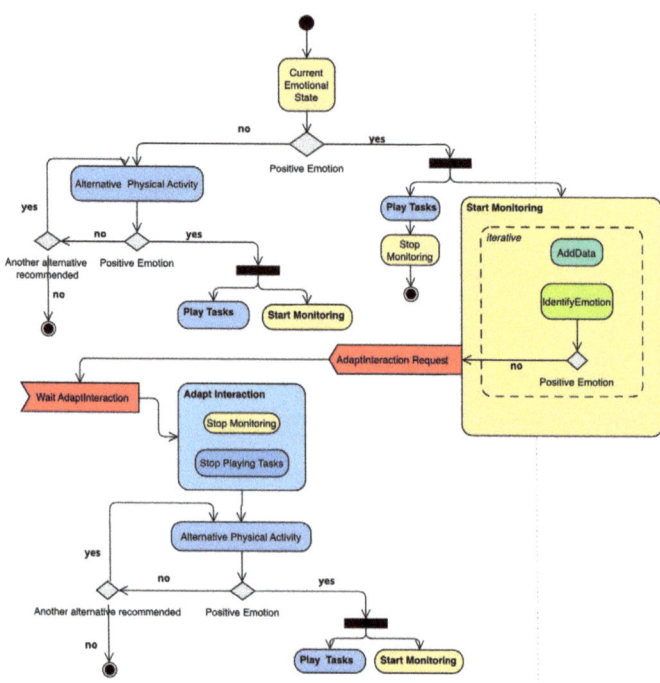

Figure 4. Activity diagram for services collaboration.

Before starting any interaction with the child, the edutainment module should call the *CurrentEmotionalState Service* from the coordinator in order to obtain the child's emotional state. If the identified emotion is positive, then the edutainment module should call the *StartMonitoring Service* from the coordinator module. When this service is called, the coordinator module should start gathering the necessary data regarding the interaction with the edutainment module (take images of the child, collect other required data), and from time to time (like 5 or 10 s) it should send this data to the *AddData Service* and/or *IdentifyEmotion Service*. If the development process is in the second phase, the gathered data should be sent only to the *AddData Service* from Database Building module. If the development process is in the third phase it should send it to the *IdentifyEmotion Service* from the emotion recognition module. In the third development phase the data could be sent to both services if we want the keep adding data to the existing datasets for further analysis and use. After receiving the identified emotion from the *IdentifyEmotion Service*, the coordinator must decide if the obtained result requires interaction adaptation of the edutainment module (for example, if the emotional state of the child is a negative one, like frustration or boredom). If it does, then the coordinator module will call the *AdaptInteraction Service* from the edutainment module. When the *AdaptInteraction Service* is called with the identified emotion, the interaction flow running in the edutainment module should be modified according to the current emotional state of the child. One possible approach for adapting the interaction flow is described in more details in the next section.

3.4. Adapting the Interaction Flow

An important aspect that must be considered for the next generation of edutainment applications is how should the interaction flow of the edutainment module be modified based on the information received from the emotion recognition module even in the classic interaction flow of an edutainment application challenges are encountered and some decisions must be taken. For example, when a child does not succeed in accomplishing a task, what should the edutainment application do? Should it show hints and how many? Should it let the child try again, and how many times? Should it just move on to the next task? Usually, the answers to these questions are found by discussing with teachers and educators, and, very often, empirical decisions are taken. In our previous work about edutainment applications development for preschool children, we have decided, together with a kindergarten teacher, that if a child gives a wrong answer to a task requirement, the application should provide hints [36,37]. However, in some cases, the decisions to be taken could be more complex. For example, when a child fails to accomplish a task, by giving the wrong answer, different situations may appear after receiving a hint:

- the child successfully accomplishes the task;
- the child fails again to perform the task;
- the child does not perform any interaction action (maybe the child leaves the computer, abandoning the interaction altogether).

For our previously developed edutainment applications, the approach used was to go on to the next task if a child failed twice to accomplish the current task. Afterwards, we considered presenting again the failed tasks to the child if he/she desires to give it another try. Furthermore, if a child does not interact with the edutainment application for 45 s, then the application automatically goes to the next task. However, in this situation, it is also possible that the child has already moved away from the computer and, in such case, the interaction should not continue.

In an emotion-aware edutainment application, decisions based on the identified emotions regarding the interaction flow must also be taken. In this scenario, it is very important to have a high accuracy of emotion recognition, but also a real-time answer regarding the emotion identification. A solution that decides the tasks' flow based on a child's emotional state would provide a better approach to support socio-emotional learning and, also, to support young children's progress in learning.

The identification of negative emotions such as frustration, anger or boredom triggers the interaction flow adaption decisions. Interaction with edutainment applications should only occur when children are in a positive emotional state. As a consequence, an emotion aware edutainment should assess children's emotions, and should play the normal interaction flow only if the child is in a positive emotional state. Otherwise, the edutainment application should suggest activities that will improve the child's emotional state by sending encouraging messages, kind messages, or suggesting fun physical activities (for example, imitate birds' flying) or relaxing activities (take deep breaths). In the following, we will call *timeout* the interruption of the interaction flow by proposing entertaining physical activities.

After the child is ready to begin the learning process (meaning that the identified emotion is not angry or frustrated), the application should start presenting the content (usually designed as a story or a game with tasks presented as challenges). If the child's emotional state does not change, then the normal interaction flow will be presented. If the child's emotional state changes, then an analysis of the situation is performed as follows. If the child becomes angry, then the interaction should stop and a timeout should be given to the child to overcome the anger. The application should propose some relaxing physical activities and the interaction should start again only if the child is in a positive mood. If the child becomes bored, then the application should switch to more complex tasks or it should change the objects in the task context or it should completely change the task type. If the child becomes frustrated, the application should try to identify the cause of frustration. The cause of frustration might be physical, in the sense that the young child cannot perform the required interaction task (for example, a drag and drop action). In this case, the application should switch to simpler interaction methods (like using only clicks, for example). If a child is frustrated because she/he does not know how to solve the task (for example, when the child does not perform any action on the interface), then some cues should be presented to support task accomplishment. If the child still cannot solve the task, the application should skip the current task and it should continue the interaction flow with encouraging messages. Still, it is difficult to automatically identify the frustration cause, but if the child continues interacting with the interface we may suppose that he/she intends to solve the task, but cannot perform the required actions.

If a child needs more than a predefined number of timeouts during the interaction, we consider that the interaction flow should stop to avoid amplification of the child's negative emotions. The experts consider that if the child requires more than two timeouts during the interaction, then the child's emotional state is not suitable for learning.

The solution that we propose for the interaction flow adaptation consists of two parts: an adaptation of the algorithm corresponding to the normal interaction flow and a newly added algorithm for when the *AdaptInteraction Service* is called by the coordinator. In Algorithm 1, the outline of the normal interaction flow algorithm that was modified in order to consider the emotional state of the child when the application starts is given.

Before starting any interaction, the algorithm gets the child's emotional state from the coordinator. If the state is a negative one, namely, angry, the application will give the child a timeout and propose some entertainment activities in order to change the child's emotional state. After the execution of the entertainment activities, the emotional state is obtained again. If it is still a negative one, the previous steps are executed again; otherwise, the normal interaction flow will start. Before starting this flow, the emotional state monitoring activity is started by calling the *StartMonitoring Service* from the coordinator. If after executing the relaxing physical activities for a predefined number of times (called *limit*), the state of the child is still a negative one, then the interaction stops. We consider that, in this case, the child's emotional state does not facilitate learning.

Algorithm 1: Normal interaction flow algorithm.

Data: *limit*-maximum number of allowed *timeOuts*
timeOuts ← 0
currentState ← @ Get CurrentEmotionalState
while *currentState = angry and timeOuts < limit* **do**
　　@Increase *timeOuts*
　　@Propose and execute some entertaining activities
　　currentState ← @ Get CurrentEmotionalState
end
if *timeOuts ≥ limit* **then**
　　@Stop interaction
else
　　@Start monitoring activity
　　while ∃ *tasks to be executed* **do**
　　　　currentTask ← @Next task
　　　　@Play *currentTask*
　　end
　　@Stop monitoring activity
end

In Algorithm 2, an outline of the algorithm that is executed each time the service *AdaptInteraction* is called by the coordinator is given. The coordinator will call this service when a negative emotion of interest is identified by the emotion recognition module, during the monitoring activity. The experts consider frustration and boredom as negative emotions from learning point of view, so the interaction flow will be adapted only when these emotions are identified by the emotion recognition module. Whenever the coordinator module gets a negative emotion of interest while in monitoring mode, the coordinator module calls *AdaptInteraction* service. The edutainment module, when *AdaptInteraction* service is called, checks the negative emotion kind, and based on this, will decide how to adapt the interaction:

- If the emotional state is frustration, then the edutainment application will resume executing the interaction flow, will stop the monitoring activity, will increase the number of *timeOuts* given and will propose relaxing physical tasks in order to change the child's emotional state. If after the execution of these tasks, the emotional state of the child becomes positive, then the monitoring activity will restart and the edutainment module will continue playing the *currentTask*. Otherwise, the interaction is stopped.
- If the emotional state is boredom, then the edutainment application will increase the difficulty level or the type of the next task to be executed.

Algorithm 2: Interaction adaptation algorithm.

Data: *state*-the emotional state of the child
currentState ← *state*
if *currentState* = *frustration* **then**
    while *currentState* = *frustration* and *timeOuts* < *limit* **do**
        @Increase *timeOuts*
        @Resume playing *currentTask*
        @Stop monitoring activity
        @Propose relaxing physical tasks
        currentState ← @ Get CurrentEmotionalState
    end
    if *currentState is positive* **then**
        @Start monitoring activity
        @Continue playing *currentTask*
    else
        @Stop interaction
    end
else
    if *currentState* = *bored* **then**
        @Change Next Task (increase difficulty, change task type, etc.)
    end
end

---

4. Prototype

In order to test our proposal for the next generation of edutainment applications, we have conducted a preliminary study in which we have modified a simple edutainment application based on the results of a facial expression emotion recognizer. A proof of concept prototype application has been developed by a team of computer science master students. The prototype adds an effect to a selected edutainment material based on the automatically identified emotions of the viewer. In this study, we have focused on recognizing the following emotions: happiness, sadness, anger, disgust and surprise, and we have applied the following effects when one of the interested emotions is identified:

- a bright effect when happiness is identified;
- a sepia effect when sadness is identified;
- a distorted effect when anger is identified;
- a blurred effect when disgust is identified;
- a black and white effect when surprise is identified.

In our investigation, we have started with two intelligent models based on Deep Learning for detecting emotions from adult faces. The first model we have tested was organised in a 6-layered Convolutional Neural Network, while the second model was organised in a 17-layered Convolutional Neural Network (both implemented in the Keras framework). These models were trained on small (48 × 48 px) grayscale images from Facial Expression Recognition (FER) dataset [32]. In this dataset, there are 28,709 training and 3589 testing images. Each of the images were stored in 48 × 48 pixels. Unfortunately, this dataset does not contain any children images.

The learning settings were characterised by a different number of epochs, Adam, derived from Adaptive Moment Estimation [38], as an optimization algorithm and a categorical cross-entropy loss function. For weight initialization we used the default settings, each layer having its own default value for initializing the weights but for most of the layers, the default kernel initializer was the Glorot uniform initialiser [39].

In our use case, we were interested in getting emotion recognition as accurately as possible, the most important criterion used for evaluating the models' quality was accuracy. With the best model, trained on adults' images, we got an accuracy of 84.42%. The loss evolution during the training process is depicted in Figure 5.

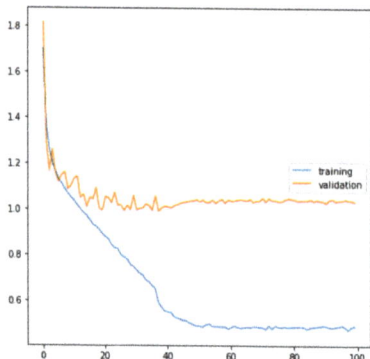

Figure 5. Loss evolution for the model trained on adults faces.

We also investigated the performance of our models in the case of children faces. The Child Affective Face Set (CAFE) [30] was used in this scope. This dataset contains 1192 photographs of 154 children. In this scenario, our second model scored the best accuracy, with an average of 75.73% (see the confusion matrix from Figure 6) and resolved some of our problems with the confusion between some of the classes. However, a liability of this model is that the loss remained on a higher value and the confusion between disgust and angry and the one between surprise and fear persisted.

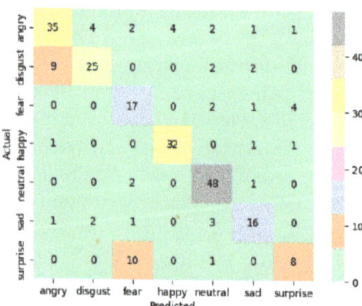

Figure 6. Real emotions versus predicted emotions in children images.

We noticed an important characteristic that influenced the recognition process in the children case, as follows. The age of the children is a factor that affects the results, possibly due to some transitions that may be much more pronounced in younger children (such as cheeks). Emotions that have similar effects on the face are confused (for example, anger and disgust which are characterized by frowning and partial or total closure of the eyes). Due to the natural shape of the face, some children are recognized as happy or sad even though they have a neutral face position.

Due to the improved accuracy of emotion recognition for adults compared to the accuracy obtained by our approach on children's images, we used adults as test subjects. In Figures 7–10 is shown how the edutainment changes based on the viewer's different emotions (anger, happy, disgust and surprise).

Figure 7. Anger emotion and distorted filter.

Figure 8. Happy emotion and bright filter.

Figure 9. Disgust emotion and blurred filter.

Figure 10. Surprise emotion and black & white filter.

5. Discussions

The results of our study show that it is feasible to develop emotions aware edutainment applications, but there are some aspects that must be improved. In the following, we discuss the advantages and disadvantages of developing such applications, and what must be further improved in order to get applications that can be used in a real context. The advantages that this type of edutainment applications bring are as follows:

- The children could safely use these applications outside the formal education system, especially when in-person interaction is not possible due to social distancing rules.
- The learning process of the child will be personalized, adapted to his/her own pace.

- Interacting with safe edutainment applications, the young child will also develop basic digital competences.
- The proposed architecture allows to easily plug-in and -out the emotion recognition and adaptation modules.

The disadvantages that using this type of edutainment applications may bring are as follows:

- Such applications could have an increased response time, affecting negatively the interaction.
- Some researchers consider that the gathered information about the children' state of mind could be improperly used to influence subconscious processes. In our proposal, the identified emotion is used only to avoid increased negative emotions while interacting with the application. If negative emotions are identified repeatedly, the application should stop executing.

Our preliminary research indicates that developing emotion-aware edutainment applications for young children is feasible, but that some improvements are required. First, the accuracy of the emotion recognizer from facial expressions must be improved. More datasets with children are needed in order to improve the existing emotions recognizers' results. Second, other sources of information for the emotion recognizer should be considered, like children's posture and motion or children's voice. Third, the time needed to identify the emotions and the time needed for adapting the interaction flow must be carefully analyzed. An aspect revealed by our study is the delayed adaptation. It takes 2–3 s until the addition of the effect is visible to the viewer. In a real-context, it shouldn't take long for the edutainment application to react to a change in the child's emotional state.

6. Conclusions and Further Work

In this paper, we have presented our proposal for the next generation of edutainment applications for young children, namely, emotion-aware edutainment applications. By enhancing edutainment applications with emotion awareness we can provide a better context for learning for young children. In the future we intend to

- validate our proposal on real and more complex case studies (implementation of the proposed approach);
- use multiple channels (body posture, voice, sensors) to extract information for the automatic emotion recognition module;
- consider the situations in which negative emotions occur frequently for different children (in this case it may also mean that changes in the design of the edutainment module should be made); and
- use emotions awareness to also evaluate the satisfaction of the little users. Identifying frustration during learning activities with an edutainment application could also provide hints on interaction flow design.

Author Contributions: Conceptualization, A.-M.G., G.-S.C. and L.-S.D.; methodology, A.-M.G., G.-S.C. and L.-S.D.; software, A.-M.G. and L.-S.D.; validation, A.-M.G. and L.-S.D.; investigation, A.-M.G., G.-S.C. and L.-S.D.; data curation, L.-S.D.; writing—original draft preparation, A.-M.G. and G.-S.C. and L.-S.D.; writing—review and editing, A.-M.G., G.-S.C. and L.-S.D.; visualization, L.-S.D.; supervision, G.-S.C.; project administration, G.-S.C. All authors have read and agreed to the published version of the manuscript.

Funding: This research received no external funding.

Institutional Review Board Statement: Not applicable.

Informed Consent Statement: Informed consent was obtained from all subjects involved in the study.

Data Availability Statement: Not applicable.

Acknowledgments: The authors would like to thank to the educational experts who provided support in understanding their work, to the students involved in case studies implementation, to the children participating in our activities and to their parents for agreeing it.

Conflicts of Interest: The authors declare no conflict of interest.

References

1. Disney, W. Educational Values in Factual Nature Pictures. *Educ. Horizons* **1954**, *33*, 82–84.
2. Rapeepisarn, K.; Wong, K.W.; Fung, C.C.; Depickere, A. Similarities and Differences between "Learn through Play" and "Edutainment". In Proceedings of the 3rd Australasian Conference on Interactive Entertainment, Perth, Australia, 4–6 December 2006; Murdoch University: Murdoch, Australia, 2006; pp. 28–32.
3. Nemec, J.; Trna, J. Edutainment or Entertainment Education Possibilities of Didactic Games in Science Education. The Evolution of Children Play-24. In Proceedings of the ICCP World Play Conference, Brno, Czech Republic, September 2007; pp. 55–64.
4. Mat Zin, H.; Mohamed Zain, N.Z. The effects of edutainment towards students' achievements. *Reg. Conf. Knowl. Integr. ICT* **2010**, *129*, 2865.
5. Kara, Y.; Yeşilyurt, S. Comparing the Impacts of Tutorial and Edutainment Software Programs on Students' Achievements, Misconceptions, and Attitudes towards Biology. *J. Sci. Educ. Technol.* **2008**, *17*, 32–41. [CrossRef]
6. Denham, S.A.; Bassett, H.H.; Thayer, S.K.; Mincic, M.S.; Sirotkin, Y.S.; Zinsser, K. Observing preschoolers' social-emotional behavior: Structure, foundations, and prediction of early school success. *J. Genet. Psychol.* **2012**, *173*, 246–278. [CrossRef]
7. Hyson, M. *The Emotional Development of Young Children: Building an Emotion-Centered Curriculum*; Teachers College Press: New York, NY, USA 2004.
8. Kostelnik, M.; Soderman, A.; Whiren, A.; Rupiper, M.L. *Guiding Children's Social Development and Learning: Theory and Skills*; Cengage Learning: Boston, MA, USA, 2016.
9. Feidakis, M. Chapter 11—A Review of Emotion-Aware Systems for e-Learning in Virtual Environments. In *Formative Assessment, Learning Data Analytics and Gamification*; Caballé, S., Clarisó, R., Eds.; Intelligent Data-Centric Systems; Academic Press: Boston, MA, USA, 2016; pp. 217–242. [CrossRef]
10. Ruiz, S.; Urretavizcaya, M.; Fernández-Castro, I.; López-Gil, J.M. Visualizing Students' Performance in the Classroom: Towards Effective F2F Interaction Modelling. In *Design for Teaching and Learning in a Networked World*; Conole, G., Klobučar, T., Rensing, C., Konert, J., Lavoué, E., Eds.; Springer International Publishing: Cham, Switzerland, 2015; pp. 630–633.
11. Druin, A.; Solomon, C. *Designing Multimedia Environments for Children: Computers, Creativity, and Kids*; Wiley: Hoboken, NJ, USA, 1996.
12. Markopoulos, P.; Bekker, M. How to compare usability testing methods with children participants. In *Interaction Design and Children*; Now Publishers Inc.: Hanover, PA, USA, 2002; Volume 2, pp. 153–158.
13. Frijda, N.H. *Appraisal and Beyond: The Issue of Cognitive Determinants of Emotion*; Lawrence Erlbaum Associates Ltd.: Hove, UK, 1993.
14. Frijda, N.H. Varieties of affect: Emotions and episodes, moods and sentiments. In *The Nature of Emotion: Fundamental Questions*; Ekman, P., Davidson, R., Eds.; Oxford University Press: New York, NY, USA, 1994; pp. 59–64.
15. Davou, B. *Thought Processes in the Era of Information: Issues on Cognitive Psychology and Communication*; Papazissis Publishers: Athens, Greece, 2000.
16. Damasio, A.R. *Descartes' Error. Emotion, Reason and the Human Brain*; Avon Books: New York, NY, USA, 1994.
17. Ekman, P.; Friesen, W. *Facial Action Coding System: A Technique for the Measurement of Facial Movement*; Consulting Psychologists Press: Palo Alto, CA, USA, 1978.
18. Ortony, A.; Clore, G.L.; Collins, A. *The Cognitive Structure of Emotions*; Cambridge University Press: Cambridge, UK, 1988. [CrossRef]
19. Parrott, W.G. *Emotions in Social Psychology: Key Readings*; Psychology Press: Oxfordshire, UK, 2000.
20. Pekrun, R. The Impact of Emotions on Learning and Achievement: Towards a Theory of Cognitive/Motivational Mediators. *Appl. Psychol.* **1992**, *41*, 359–376. [CrossRef]
21. Pekrun, R.; Lichtenfeld, S.; Marsh, H.W.; Murayama, K.; Goetz, T. Achievement Emotions and Academic Performance: Longitudinal Models of Reciprocal Effects. *Child Dev.* **2017**, *88*, 1653–1670. [CrossRef]
22. Rowe, A.D.; Fitness, J. Understanding the Role of Negative Emotions in Adult Learning and Achievement: A Social Functional Perspective. *Behav. Sci.* **2018**, *8*, 27. [CrossRef]
23. Manwaring, K.C. Emotional and Cognitive Engagement in Higher Education Classrooms. Ph.D. Thesis, Brigham Young University, Provo, UT, USA, 2017. Available online: https://scholarsarchive.byu.edu/etd/6636 (accessed on 22 December 2021).
24. Halberstadt, A.G.; Eaton, K.L. A meta-analysis of family expressiveness and children's emotion expressiveness and understanding. *Marriage Fam. Rev.* **2002**, *34*, 35–62. [CrossRef]
25. Taskiran, M.; Kahraman, N.; Erdem, C.E. Face recognition: Past, present and future (a review). *Digit. Signal Process.* **2020**, *106*, 102809. [CrossRef]
26. Adjabi, I.; Ouahabi, A.; Benzaoui, A.; Taleb-Ahmed, A. Past, Present, and Future of Face Recognition: A Review. *Electronics* **2020**, *9*, 1188. [CrossRef]

27. Deng, J.; Guo, J.; Zhang, D.; Deng, Y.; Lu, X.; Shi, S. Lightweight face recognition challenge. In Proceedings of the IEEE International Conference on Computer Vision Workshops, Seoul, Korea, 27–28 October 2019
28. Ko, B.C. A brief review of facial emotion recognition based on visual information. *Sensors* **2018**, *18*, 401. [CrossRef]
29. Lopes, A.T.; de Aguiar, E.; De Souza, A.F.; Oliveira-Santos, T. Facial expression recognition with convolutional neural networks: coping with few data and the training sample order. *Pattern Recognit.* **2017**, *61*, 610–628. [CrossRef]
30. LoBue, V.; Thrasher, C. The Child Affective Facial Expression (CAFE) set: Validity and reliability from untrained adults. *Front. Psychol.* **2015**, *5*, 1532. [CrossRef]
31. Lucey, P.; Cohn, J.F.; Kanade, T.; Saragih, J.; Ambadar, Z.; Matthews, I. The extended cohn-kanade dataset (ck+): A complete dataset for action unit and emotion-specified expression. In Proceedings of the 2010 IEEE Computer Society Conference on Computer Vision and Pattern Recognition-Workshops, San Francisco, CA, USA, 13–18 June 2010; IEEE: New York, NY, USA, 2010; pp. 94–101.
32. Goodfellow, I.J.; Erhan, D.; Carrier, P.L.; Courville, A.; Mirza, M.; Hamner, B.; Cukierski, W.; Tang, Y.; Thaler, D.; Lee, D.H.; et al. Challenges in representation learning: A report on three machine learning contests. In *International Conference on Neural Information Processing*; Springer Publishing: New York, NY, USA, 2013; pp. 117–124.
33. Lyons, M.; Akamatsu, S.; Kamachi, M.; Gyoba, J. Coding facial expressions with gabor wavelets. In Proceedings of the Third IEEE International Conference on Automatic Face and Gesture Recognition, Nara, Japan, 14–16 April 1998; IEEE: New York, NY, USA, 1998; pp. 200–205.
34. Guran, A.M.; Cojocar, G.S.; Diosan, L. A Step Towards Preschoolers' Satisfaction Assessment Support by Facial Expression Emotions Identification. Knowledge-Based and Intelligent Information & Engineering Systems. In Proceedings of the 24th International Conference KES-2020, Virtual Event, 16–18 September 2020. [CrossRef]
35. Goodfellow, I.; Bengio, Y.; Courville, A. *Deep Learning*; MIT Press: Cambridge, MA, USA, 2016.
36. Guran, A.M.; Cojocar, G.S.; Moldovan, A. Designing edutainment software for digital skills nurturing of preschoolers: A method proposal. In Proceedings of the ACM/IEEE 42nd International Conference on Software Engineering: Software Engineering in Society, ICSE-SEIS '20, Seoul, Korea, 27 June–19 July 2020; pp. 63–70. [CrossRef]
37. Guran, A.M.; Cojocar, G.S.; Moldovan, A. A User Centered Approach in Designing Computer Aided Assessment Applications for Preschoolers. In Proceedings of the 15th International Conference on Evaluation of Novel Approaches to Software Engineering, ENASE 2020, Prague, Czech Republic, 5–6 May 2020; pp. 506–513. [CrossRef]
38. Kingma, D.P.; Ba, J. Adam: A Method for Stochastic Optimization. In Proceedings of the 3rd International Conference for Learning Representations, Diego, CA, USA, 7–9 May 2015.
39. Glorot, X.; Bengio, Y. Understanding the difficulty of training deep feedforward neural networks. In Proceedings of the Thirteenth International Conference on Artificial Intelligence and Statistics, Sardinia, Italy, 13–15 May 2010; Chia Laguna Resort; Teh, Y.W., Titterington, M., Eds.; JMLR, Inc. and Microtome Publishing: Brookline, MA, USA, 2010; Volume 9, pp. 249–256.

Article

Improving User's Experience in Exploring Knowledge Structures: A Gamifying Approach

Brigitte Breckner [1,†,‡], Christian Săcărea [2,*,†,‡] and Raul-Robert Zavaczki [3,†,‡]

[1] Department of Mathematics, Institute for German Studies, Babeș-Bolyai University, 400084 Cluj-Napoca, Romania; brigitte.breckner@ubbcluj.ro

[2] Department of Computer Science, Institute for German Studies, Babeș-Bolyai University, 400084 Cluj-Napoca, Romania

[3] Institute for German Studies, Babeș-Bolyai University, 400084 Cluj-Napoca, Romania; raul.zavaczki@scs.ubbcluj.ro

* Correspondence: christian.sacarea@ubbcluj.ro

† Current address: Babeș-Bolyai University, Kogalniceanu 1 Str., 400084 Cluj-Napoca, Romania.

‡ These authors contributed equally to this work.

Abstract: Gamifying user experience while navigating knowledge structures is the new paradigm in online learning and improving human–computer interaction. Learning by playing and learning by doing lie at the core of this approach. Rooted in the paradigm of Conceptual Knowledge Processing and conceptual landscapes of knowledge, this research proposes an immersive experience in 3D knowledge structures, the VR FCA project. We exemplify this experience, at least the amount possible in a journal paper, by using data sets about topological spaces, a mathematical field which is notoriously hard to grasp by beginners or sometimes even at an intermediate level. We present the stages of this project, the technologies we have used in the implementation and also some basics of Formal Concept Analysis (FCA), the mathematical theory which lies at heart of Conceptual Knowledge Processing.

Keywords: formal concept analysis; conceptual knowledge processing; virtual reality

MSC: 97P80; 68T35

Citation: Breckner, B.; Săcărea, C.; Zavaczki, R.-R. Improving User's Experience in Exploring Knowledge Structures: A Gamifying Approach. *Mathematics* **2022**, *10*, 709. https://doi.org/10.3390/math10050709

Academic Editor: Bo-Hao Chen

Received: 31 December 2021
Accepted: 18 February 2022
Published: 24 February 2022

Publisher's Note: MDPI stays neutral with regard to jurisdictional claims in published maps and institutional affiliations.

Copyright: © 2022 by the authors. Licensee MDPI, Basel, Switzerland. This article is an open access article distributed under the terms and conditions of the Creative Commons Attribution (CC BY) license (https://creativecommons.org/licenses/by/4.0/).

1. Introduction

From the very beginning of the Information Age, investigating the structure of knowledge was a central approach. Cognitive sciences, psychology and Artificial Intelligence were equally concerned about defining and understanding the structure of knowledge.

The term knowledge structure was established in the 1970s in Cognitive Sciences, a discipline that emerged at the intersection of Psychology and Computer Science [1]. Jonassen [2] defines knowledge structures in education as the knowledge of how concepts in a domain are interrelated. According to this author, these structures are utilised in intelligent educational systems to adapt navigation, to provide knowledge tracing, or learner guidance. The role of knowledge structure in learning was investigated by Ley in [3]. Here, he analyses the experience of more than 10 years of research intending to offer intelligent services through capturing and leveraging knowledge structures in workplace learning.

Formal Concept Analysis (FCA) is a prominent field of applied mathematics providing an elegant, intuitive yet powerful knowledge representation as order diagrams. Using the conceptual landscapes of knowledge paradigm of R. Wille [4], knowledge is activated by using the landscape metaphor and FCA and its algorithmic machinery plays an important role in mining knowledge, identifying patterns, processing concepts and representing them. The basic data set of FCA is the formal context, i.e., a collection of objects and attributes linked by a binary relation, called incidence relation. This data set is usually represented as a two-dimensional data table, encoding in 2D the entire information contained in the data set. Formal concepts are the mathematisation of the traditional understanding of a concept, i.e., a maximal collection of objects and commonly shared attributes. In his seminal paper, R. Wille [5] proposes an order theoretical approach to the problem of concept mining. More precisely, using a Galois connection between the power sets of the sets of objects and that of attributes, concept forming operators are defined. The Basic Theorem on Concept Lattices [6] proves that, for each formal context, the correspondent concept set carries a complete lattice structure and each complete lattice can be obtained as a concept lattice of some formal context.

This was the starting point of the development of a brand new research field, with major applications in Artificial Intelligence, Knowledge Processing and Representation. The concept lattice or concept hierarchy represents the concepts, i.e., knowledge units and their hierarchy, and builds the basis for further communication, navigation and investigation of the data set. Conceptual structures gained more and more importance and they became the counterpart of knowledge structures from the cognitive sciences in FCA. Reference [7] provides an overview on how FCA and Conceptual Graphs (CGs) can be used in many areas, serving as a benchmark of research on conceptual structures.

This research starts from the definition of Jonassen [2] and uses a new approach to visualisation and navigation of concept sets in knowledge structure based on FCA and virtual reality (VR) following the outlines of Conceptual Knowledge Processing defined by R. Wille in [8]. According to R. Wille, "Conceptual Knowledge Processing is understood as the general scientific discipline which activates acts of thinking such as representing, reasoning, acquiring, and communicating conceptual knowledge". We prove that moving from a two-dimensional representation of conceptual structures into a three-dimensional virtual reality representation of these structure not only enhances understanding of these knowledge structures but also opens new possibilities to navigate, explore and communicate over the knowledge represented herein.

From our own experience in undertaking research projects and from the experience we gathered over the last decades from the main FCA community, there is a gap between the results, projects, algorithms and software tools developed within the FCA community and their usage outside the scientific community. While other related fields, such as, for instance, learning space theory, have software implementations which are heavily used by millions of users (https://www.aleks.com (accessed on 15 December 2021)), the usage of FCA tools and algorithms remains modest. It is not the aim of this research to discuss the reasons for this situation but to propose an approach which has the potential to close the gap between the powerful machinery and elegant knowledge representation of FCA and regular users.

We propose a gamifying approach towards exploring knowledge structures not in the sense that we transform navigation in concept lattice in a computer game but we offer users the possibility to explore 3D concept lattices in a VR environment, to rotate these structures, to fly around, to teleport from one node to another, to laser point to a certain node and to interact with these FCA based conceptual landscapes in a more lively way than on a regular 2D computer screen. This approach has been extensively tested by the FCA community in 2019 at the last prepandemic International Conference on Formal Concept Analysis. The feedback gathered from senior and junior researchers helped us to further improve the system to a multiuser VR based FCA navigation tool fulfilling R. Wille's dream of navigating conceptual landscapes of knowledge in a similar way we navigate and explore natural landscapes.

The present research presents a new use case of some previous results [9], which deepens [10] focusing on a proof of concept of how FCA enhanced with VR has the potential both to improve user's experience in exploring knowledge structures and also to gamify this approach. The multiuser modus also makes it possible to communicate in this

VR environment, to analyse the structure of knowledge and to highlight interesting parts beyond the classical knowledge representation of knowledge.

2. Formal Concept Analysis

In the following, we briefly sketch some basic notions about FCA. For more, please refer to [6].

The fundamental data structure in FCA is a so-called *formal context* which consists of two sets of objects and attributes, respectively. These sets are denoted in the literature by G and M, respectively, from the German word *Gegenstände* for objects and *Merkmale* for attributes. The fact that an object $g \in G$ has an attribute $m \in M$ is formalised by a binary relation $I \subseteq G \times M$. Hence, a formal context can be represented as a triple $\mathbb{K} = (G, M, I)$. Starting from this very basic data structure, there are many other enhancements, such as many-valued context, i.e., objects have attributes with a certain value, or triadic FCA, i.e., objects have attributes under certain conditions, etc.

For finite formal contexts, a very natural and at hand representation is by using a cross-table. Cross-tables can be easily understood by humans; the rows are labelled by object names, the columns are labelled by attribute names. The incidence relation is represented by crosses in that table (see Figure 1).

The whole point in FCA that makes the difference between other search or representation methods in data sets (SQL, rough sets, clustering, etc.) is the focus on concepts, which are understood as maximal clusters of information, bearing a clear meaning. While in philosophy the notion of a concept has been extensively discussed, FCA gives an elegant formalisation of a concept based on a Galois connection between the power sets of the object and attributes sets, respectively. This Galois connection is defined via the so-called *derivation operator*. More precisely, for a subset $A \subseteq G$ of objects, the derivation of A is defined as the set of all commonly shared attributes of all objects $g \in A$, $A^I := \{m \mid gIm \text{ for all } g \in A\}$. Dually, the derivation of a subset $B \subseteq M$ of attributes is defined as the set of all objects sharing exactly the attributes in B, $B^I = \{g \mid gIm \text{ for all } m \in B\}$.

The derivation operators and the subsequent closure systems that this Galois connection gives rise can be seen as *concept forming operators*. By this, we can define a *formal concept* as being a pair (A, B), where $A \subseteq G$ and $B \subseteq M$ are satisfying $A^I = B$ and $B^I = A$. This is similar to a clustering process, with the condition that the clusters are maximal with respect to the incidence relation. In the cross-table representation, formal concepts are maximal rectangles of crosses obtained by permuting rows and columns.

If (A, B) is a formal concept, the set A is called *extent* and B is called *intent*. The set of all concepts of a formal context $\mathbb{K} = (G, M, I)$ is denoted by $\mathfrak{B}(\mathbb{K})$.

The set $\mathfrak{B}(\mathbb{K})$ is naturally ordered by the subconcept–superconcept relationship. We say that (A, B) is a *subconcept* of (C, D) (or equivalently, (C, D) is a *superconcept* of (A, B)), and we write $(A, B) \leq (C, D)$ if and only if $A \subseteq C$ ($\Leftrightarrow D \subseteq B$). The fundamental theorem of FCA states that $(\mathfrak{B}(\mathbb{K}), \leq)$ is a complete lattice and every complete lattice is obtained as the concept lattice of a suitable chosen formal context. The graphical representation of this conceptual hierarchy can be achieved in an elegant way by using *order diagrams*.

Figure 2 shows the concept system of the context in Figure 1. In this figure, we considered the reduced labelling representation generally used in FCA to improve readability: Each node is a concept but only the *object concepts* and *attribute concepts* are labelled. These concepts are particular concepts of the form (g'', g') and (m', m'') for $g \in G$ and $m \in M$. Since some of these object or attribute concepts can overlap, i.e., some objects and attributes can generate the same object concept or attribute concept, respectively, it might happen that some of these concepts bear more than one object or attribute label. Due to the above mentioned Fundamental Theorem of FCA, every other concept is obtained as the infimum or supremum of object concepts or attribute concepts.

	needs water to live	lives in water	lives on land	needs chlorophyll	dicotyledon	monocotyledon	can move	has limbs	breast feeds
fish leech	×	×					×		
bream	×	×					×	×	
frog	×	×	×				×	×	
dog	×		×				×	×	×
water weeds	×	×		×		×			
reed	×	×	×	×		×			
bean	×		×	×	×				
corn	×		×	×		×			

Figure 1. Example of a formal context to data set Living Beings and Water.

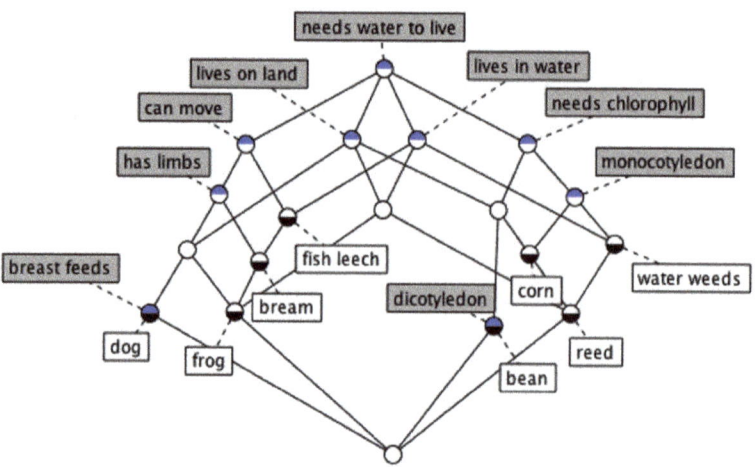

Figure 2. Concept lattice of the formal context in Figure 1.

This also explains how a concept lattice should be read, respectively how navigation in concept lattices takes place: Each node is a concept. The extent is obtained by collecting all object names which we find as labels for concepts which are reachable by downward going paths. Dually, the intent is obtained by collecting all attribute names which are reachable by upward going paths. The upmost node contains all attributes which are commonly shared by *all objects* in the formal context, i.e., full columns in the cross-table representation. Dually, the bottommost concept contains all objects which have *all attributes* of the context, i.e., full rows in the cross-table representation.

3. Setting the Stage: The VR FCA Project

From its very beginnings in the late '80s of the previous century, FCA offered a very elegant way to represent knowledge structures as order diagrams and the aesthetics of these diagrams were always valued in R. Wille's group. Over the decades, many FCA graphical representation tools have appeared. A very early project was ConImp [11] which was devoted to the computation of implications and implementation of some early algorithms.

A meanwhile classic tool for dyadic FCA is Concept Explorer (conexp.sourceforge.net). ConExp-FX (https://github.com/francesco-kriegel/conexp-fx (accessed on 15 December 2021)), and ConExp-NG (https://github.com/fcatools/conexp-ng (accessed on 15 December 2021)) are reimplementations of the former. The latter one is part of FCA Tools, a collection of FCA related software tools (https://github.com/fcatools (accessed on 15 December 2021)). An overview of attribute exploration with domain knowledge was given in [12]. FCA Tools Bundle [13] is an ambitious project aiming to gather all FCA Tools in a single software over a web interface. Besides the classical dyadic approach, FCA Tools Bundle handles tricontexts, navigation tools in triadic data sets, contains an implementation of a local navigation paradigm in triadic data sets as well as a like–dislike Answer Set Programming based method to narrow down the set of concepts we would like to investigate, and many more. LatViz [14] is a joint effort of another FCA research community to improve existing tools, implementing new functionalities such as visualisation of Pattern Structures, concept annotations, intuitive visualisation of implications, etc.

The VR FCA project [9,10] is an attempt to include modern graphic capabilities, new technologies and game engines in FCA software tools. The aim is gamifying the FCA experience in order to activate the Conceptual Knowledge Processing features necessary for a good learning experience. For this, we need an intuitive and captivating technological environment which triggers a new kind of navigating conceptual structures experience. Once you enter a VR room, a 3D concept lattice unfolds in front of you, allowing you to interact with it, to fly around, to rotate, to teleport or to narrow the perspective to a certain group of nodes. The multiuser experience allows users to simultaneously enter the same VR room and to discuss and interact all together. This research is based on [9]. Here, we focus on improvements and applications in particular fields, such as Topology, in order to highlight that gamifying the FCA experience is a major step forward in designing new intelligent systems aiming for a better human–computer interaction.

One of the major challenges was how to represent a concept lattice in three dimensions. Diagrams, and particularly order diagrams, are used in Knowledge Representation in order to reduce the cognitive load for the user. In this research, we focus on a particular representation of conceptual hierarchy as an order diagram. Nevertheless, there is no canonical representation of an order diagram in 2D, which makes it even more difficult to represent if in 3D. Following the school of R. Wille, the founder of FCA, which emphasises the need of *nice order diagrams* to support the user in his knowledge structure navigation, 2D order diagrams as many parallelograms as possible and avoids overlapping of the nodes and unnecessary edge intersections. However, this still does not solve the problem of 3D representation of a conceptual hierarchy.

A visualisation method which proved to be very effective was described in [15] and is called *3D Cone Tree*. A cone tree is a hierarchical knowledge representation laid out uniformly in three dimensions, also using the depth of the monitor not only the height or width of the monitor. In this seminal paper, cone trees rotate whenever a node is selected, and each node in the path from the selected node up to the top are brought to the front and highlighted. The rotations follow the shortest path, and they are animated. By this, the user can see the transformation and is able to better understand the underlying structure of the represented knowledge system. Paper [15] proposes interactive animation is used to shift some of the user's cognitive load to the human perceptual system. Instead of this, our approach is using Unity and SteamVR technologies to enable interactivity with the graphical representation of knowledge structures.

This research is proposing a particular approach to the exploration of knowledge structures by moving the entire experience into a 3D VR environment and gamifying this experience by using the capabilities of state-of-the-art VR technologies. By this, the idea of [15] of using 3D representations of knowledge structures is realised at a completely other level the authors thought might be possible in their paper. Even more, it also brings into life the paradigm of R. Wille of considering concept hierarchies of knowledge structures as conceptual landscapes of knowledge and to explore and navigate them according to the paradigm of landscape.

Upgrading the navigation experience to 3D and using modern VR technologies allows us to shift the perception and also the cognitive experience of navigating knowledge structures to another level. As experiments showed, this might be quite an overwhelming experience, but the gaming experience the user has by rotating, flying around, teleport or laser pointing concepts does the trick.

As mentioned in [9], the FCA VR tool is designed specifically for people that had no connection to FCA before. In order to underline the gamifying experience of navigating 3D knowledge structures, we propose in this research a use case related to concept hierarchies built over some abstract topological notions. From our own teaching experience, they are notoriously hard to grasp for beginners or even mathematicians with less background in Topology.

This idea is by itself a real challenge and this challenge is doubled by the fact that every user is different and has another understanding of how a graphical representation should look in order to match their expectations of a working system. To overcome this problem, we gathered feedback from various sources, from students, junior researchers, professors teaching at the University and active researchers in the FCA community. Because feedback came from such many places, it gave us a good overview over the overall aspect of the application, the common ground of these people, and what could be improved in the future.

Without entering into details which can be found in [9], the VR FCA project is using Unity as the engine on which we run the applications. Because it offers a scripting API over the engine, and you can build a *virtual environment* that can be later rendered to the user, it has been proved to be of a huge value since we did not have to use low-level graphics rendering libraries such as OpenGL.

SteamVR is a plugin for Unity which allows virtual reality application developers to target one single API that all the popular virtual reality headsets can connect to. HoverUI Kit is a tool that allows the developer to create virtual reality interfaces, based on the mechanism of hovering.

The use of Unity technologies with the SteamVR plugin had very many deciding factors:

- Unity with the SteamVR plugin took the responsibility of presenting the objects in the VR world, allowing us to focus on functionality of the application rather than writing our own.
- The SteamVR API is simply to use, while learning Unity is also simple since everything is based on C#.
- At the time of writing the application, the SteamVR was the best well-maintained VR Abstraction framework we could find.
- SteamVR provided better Input Controls Configuration than other plugins.

4. Describing the Stage: App Description

The core of the presentation layer is the MVC model, where we synchronise the user and multiplayer input to actual models and views that can be shown to the user. Figure 3 gives an overview of the VR FCA Application. For details about various VR related functionalities, we refer to https://github.com/zraul123/FCAVR and [9,10] (accessed on 15 December 2021).

This presentation module has the role of abstracting all inputs and data into a form presentable to the user. Additionally, here we have wrappers over Unity objects and UI elements—the UnityObjectFactory and Menus modules.

The UnityObjectFactory wrapper takes simply a View and creates its 3D Lattice representation with the data it is given.

The Menus wrapper contains all the Menus available (Main Menu, Tutorials).

The Scenario module serves as an abstraction layer over the various lattices we support—regular concept lattices, conceptual scales and Lifetrack lattices (for Temporal Concept Analysis, see [10]). The Scenario module uses the FileSystem Repository to extract the lattices definitions (we have a predetermined path for the lattice locations), which are then presented to the user as available lattices. If a user selects a lattice, it will load in the SQLite Repository its metadata, and depending on the Lattice type, it may also store the objects there (for conceptual scales and Lifetrack lattices) for quicker reference. The lattice definition is loaded from the filesystem, and the file is parsed using the LatticeService, which is actually a business wrapper over Formal Concept Analysis algorithms. The FCALib is simply a module in which we keep the Formal Concept Analysis algorithms we need (currently, only the parsing of the file format).

The MultiplayerSync modules serves as a facade for all network operations transmitted. This may be saving of nodes or additional metadata. It opens an internal WebSocket port, which is used for bi-directional communication with the server. This also contains code for Voice communication.

The Multiplayer Server external application serves as a router and broadcaster for websockets. It registers sessions/rooms and makes sure everyone in a particular session receives the network packages. This application should, obviously, live on an open port (using port-forwarding rules, for example) that is able to listen/send data to the external network.

The Multiplayer Configuration external application is an internal application to the user. It is used as a sidecar for the VR application to allow the user to configure externally to which server he wishes to connect to. The reason this was not performed inside the VR application is that it is challenging and tedious to type in VR.

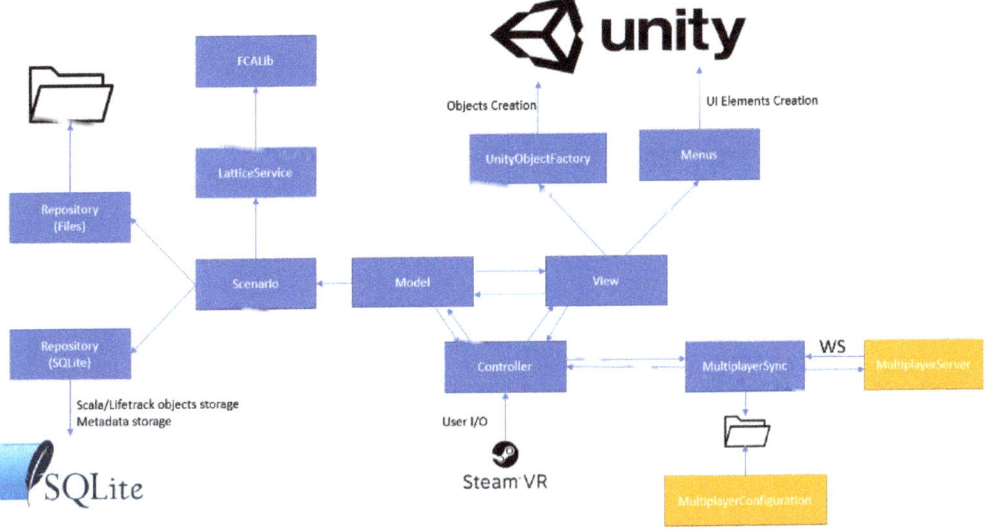

Figure 3. Structure of the VR FCA Application.

The lattice format is parsed, providing a list of nodes with their subconcept-superconcept relations. Then, we calculate the longest node-chain we can find, which gives us the depth of the lattice. Having the depth of the lattice, we can create at each increment $(1, 2, 3, \ldots, n)$ a circle around the Y axis. This circle will serve as a reference for position calculation later.

After all circles are created, we assign each node to its corresponding circle, based on its depth.

In this last step, depending on how many nodes we have on a circle, we have a step parameter that distributes nodes uniformly around its 360 angle. After this is achieved, we can simply calculate the nodes position relative to the circle centre (which, in this case, is the lattice centre on that plane) using sin and cos.

The module responsible for calculating the node position and move is the Move module, by using the following Algorithm 1:

Algorithm 1 Node position calculation.

1: max_chain_length = the maximum length of a chain
2: **for** current_length in max_chain_length **do**
3:     depth_nodes = find the nodes with depth current_length
4:     angle_deg = 20
5:     width_scale_factor = Math.clamp(length(depth_nodes), 1, 10) / 4 + 1
6:     **for** node in depth_nodes **do**
7:         node.position.x = sin(angle) + circle_base_with * width_scale_factor
8:         node.position.y = cos(angle) + circle_base_with * width_scale_factor
9:     angle += (360 / length(depth_nodes))
---

5. VR Enhanced Navigation at Work

We exemplify the capabilities of the VR FCA project in navigating complex data sets, using a more unusual data set, namely various properties of topological spaces. The data have been gathered from [16], and we refer to that work for a complete list of topologies and properties. From our experience, both as university staff and students, teaching mathematical notions can be sometimes difficult because of their abstractness and lack of visualisation. Moreover, the connections between various notions and concepts described in mathematics textbooks are hard to grasp for a beginner or even at an intermediate level and there is a constant need for study and practice in order to achieve mastery. Following the idea of gamifying the human–computer interaction and enhance it with new capabilities offered by VR technologies (such as fly, teleport, laser pinpoint, rotate, etc.), we have selected three different data sets from [16].

The first data set is the so-called Connectedness Chart, exemplifying various connectedness properties in topological spaces. The attributes are various connectedness properties, while the objects are topologies.

C_1: connected C_2: path connected C_3: arc connected
C_4: hyperconnected C_5: ultraconnected C_6: locally connected
C_7: locally path connected C_8: locally arc connected
C_9: biconnected.

The 3D visualisation of the knowledge structure of Table 1 is given in Figure 4.

While representing these 3D diagrams in 2D is somehow misleading for the entire immersive VR experience, short movies showing only rotations of these structures can be seen at https://www.cs.ubbcluj.ro/~fca/paper/ (accessed on 15 December 2021).

Teleporting, laser pointing, floating and flying through these structures, together with many-players experience, contribute to the deeper understanding of these structures and of the topic under study.

Let us explain briefly how all these features work. A user may be able to connect to a remote or local *virtual room*. When a virtual room is first launched up, the creator of the virtual room needs to select a conceptual structure that they want to inspect. After this is performed, any user can connect to this virtual room. All actions that can be performed, such as lattice rotating, node moving, scale are mapped to the controller. In the following, we have included an example for the controller from the HTC Vive headset, which we used for testing and experimenting (see Figure 5, where the footnote refers to the browsing capabilities of the system through the conceptual scales of lifetrack lattices).

All actions are handled on the client side. This means that, if you rotate or move the nodes around, this will be visible only for yourself. This has been conducted for multiple reasons:

- Performance wise, we have far less data to send/receive and handle.
- The user is free to explore without any distraction or constraint (such as being in the same *stage* of conceptual scale lattices).
- Solves most concurrency issues that we may run into.

Table 1. Formal context of Connectedness Properties.

	C_1	C_2	C_3	C_4	C_5	C_6	C_7	C_8	C_9
G_2						x	x	x	
G_6	x	x		x		x	x		x
G_9	x	x			x	x	x		x
G_{11}						x	x		
G_{12}	x			x		x	x		
G_{13}	x	x	x	x		x	x	x	
G_{20}	x	x	x			x	x	x	
G_{23}	x								x
G_{27}	x					x			
G_{29}	x					x			
G_{32}	x	x	x	x	x	x	x	x	
G_{33}	x	x	x	x		x	x	x	
G_{35}						x			
G_{36}	x	x		x		x	x		
G_{37}	x	x			x	x	x		
G_{39}	x								
G_{40}	x					x			
G_{43}	x								
G_{47}	x								
G_{70}	x	x	x						
G_{78}	x								
G_{79}	x	x	x						
G_{80}	x								
G_{81}	x	x	x						
G_{82}	x	x			x				
G_{83}	x								x
G_{84}	x								x
G_{85}	x	x	x			x	x	x	

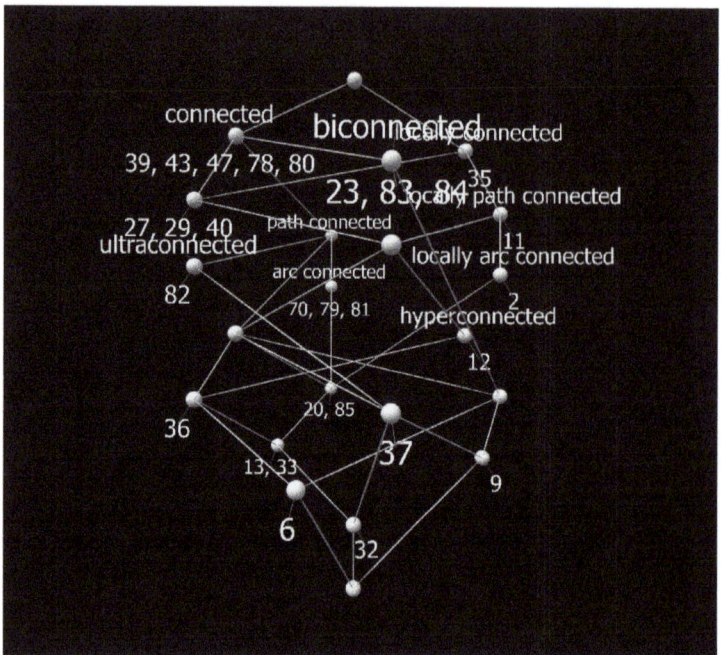

Figure 4. Three-dimensional concept lattice for the connectedness chart.

Controller Button	Left Controller	Right Controller
Trigger press	Move	Rotate
Grip short press	Reset position/rotation	Select node
Grip long press	Reset position/rotation	Next/Previous Diagram*
Touchpad press	Save node	-
Touchpad scroll	-	Zoom

* - only applicable to Scale/Lifetrack lattices

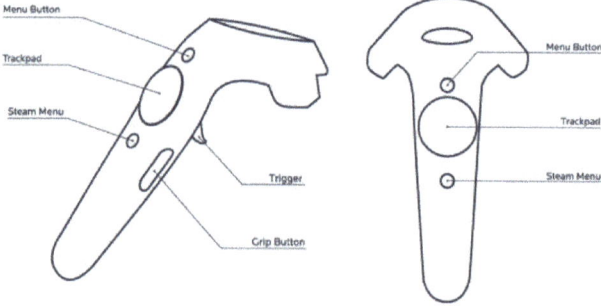

Figure 5. HTC Vive controls for navigating knowledge structures.

For the multiplayer version, the question is what is shared between the players so that we can really call it multiplayer?

- Users have a voice chat that allows them to speak directly from the VR environment.
- Users can save nodes and map them to different colours, for easier identification. Saving a node also puts it into a shared menu, where other users can select a saved node and the node will be highlighted and focused.

We briefly include two more examples of topological properties and knowledge structures. The next one is the separation axioms chart. These properties play a major role in Topology, since they clarify, for instance, whether a net is convergent or not. For details about separation axioms and their definitions, see [16] or [17].

The 3D concept lattice of the formal context considered in Table 2 is given in Figure 6.

Compact sets play a central role in various fields of Mathematics. Different compactness properties are mentioned in [16].

The corresponding compactness properties chart is given by Table 3.

The compactness 3D chart in Figure 7 is quite complex and, with a bit of imagination, an immersive 3D VR experience of teleporting and flying through this structure can be imagined. For more, we refer to the above mentioned link.

Table 2. Formal context of separation properties.

	A_1	A_2	A_3	A_4	A_5	A_6	A_7	A_8	A_9	A_{10}	A_{11}	A_{12}	A_{13}	A_{14}	A_{15}
O_1					×	×	×	×							
O_2	×														
O_3	×						×	×							
O_4	×	×													
O_5	×	×	×	×	×	×	×	×	×	×	×	×	×	×	
O_6	×	×	×	×	×	×	×	×	×	×	×	×	×	×	×
O_7							×	×							
O_8															
O_9	×							×							
O_{10}	×	×	×						×						
O_{11}	×	×	×	×								×			
O_{12}	×	×	×												
O_{13}	×	×	×	×					×						
O_{14}	×	×	×	×	×	×			×	×	×	×			
O_{15}	×	×	×	×	×	×	×		×	×	×	×	×		
O_{16}	×	×	×	×					×		×				
O_{17}	×	×	×	×	×				×	×					
O_{18}	×	×	×	×	×				×	×	×				
O_{19}	×	×	×	×											

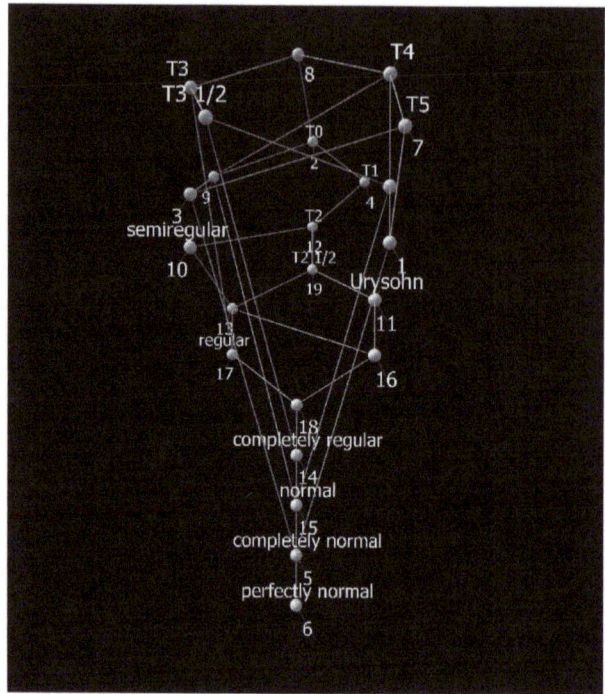

Figure 6. Separation axiom chart: concept lattice of the formal context in Table 2.

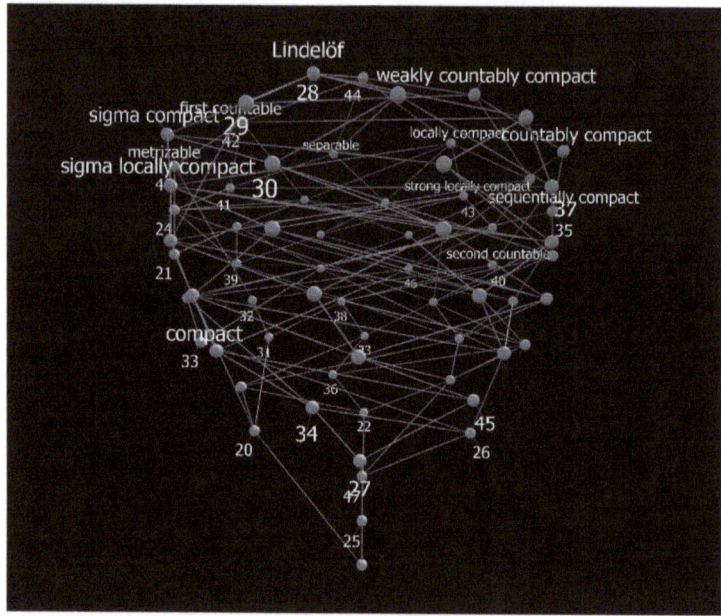

Figure 7. Compactness chart: concept lattice of the formal context in Table 3.

Table 3. Formal context of compactness properties.

	A_{16}	A_{17}	A_{18}	A_{19}	A_{20}	A_{21}	A_{22}	A_{23}	A_{24}	A_{25}	A_{26}	A_{27}	A_{28}
O_{20}		x	x				x	x	x	x	x	x	x
O_{21}							x	x				x	x
O_{22}		x	x			x	x	x	x	x	x	x	
O_{23}		x	x				x		x	x	x	x	
O_{24}							x			x		x	
O_{25}	x	x	x	x	x	x	x	x	x	x	x	x	
O_{26}	x	x	x	x	x	x	x	x	x		x		
O_{27}	x	x	x	x	x	x	x	x	x	x			
O_{28}			x										
O_{29}			x		x								
O_{30}		x	x							x			
O_{31}		x	x							x	x	x	x
O_{32}			x							x	x	x	x
O_{33}				x	x	x	x	x				x	
O_{34}	x	x	x	x	x	x	x	x	x				
O_{35}				x	x	x							
O_{36}		x	x			x	x		x	x	x	x	
O_{37}			x							x		x	
O_{38}		x	x							x	x	x	
O_{39}							x	x		x		x	
O_{40}			x							x	x	x	
O_{41}										x		x	
O_{42}												x	
O_{43}							x	x					
O_{44}													
O_{45}	x	x	x	x		x	x	x	x				
O_{46}				x		x	x	x					
O_{47}	x	x	x	x	x	x	x	x	x	x		x	
O_{48}												x	x

6. Conclusions

In this paper, we have shown a design and implementation plan of a virtual reality tool that makes possible a VR immersive experience of 3D representation of knowledge structure, aiming a gamification of this experience.

As future work, we intend to extend the application in various ways, but the most important extensions are:

- Include attribute and concept exploration strategies.
- Visualisation Algorithms: Using genetic algorithms, the visualisation of the lattice can be greatly improved.
- Navigation Assistant: Defining what the user is looking for, an AI assistant could be helping the user find the desired node.

Author Contributions: Conceptualisation, C.S. and B.B.; methodology, C.S.; software, R.-R.Z.; validation, C.S., B.B. and R.-R.Z.; formal analysis, B.B. All authors have read and agreed to the published version of the manuscript.

Funding: The publication of this article was supported by the 2021 Development Fund of the UBB.

Institutional Review Board Statement: Not applicable.

Informed Consent Statement: Not applicable.

Data Availability Statement: Not applicable.

Conflicts of Interest: The authors declare no conflict of interest.

References

1. Schank, R.C.; Abelson, R.P. *Scripts, Plans, Goals, and Understanding: An Inquiry Into Human Knowledge Structures*; Psychology Press: London, UK, 2013.
2. Jonassen, D.H.; Wang, S. Acquiring Structural Knowledge from Semantically Structured Hypertext. Available online: https://eric.ed.gov/?id=ED348000 (accessed on 15 December 2021).
3. Ley, T. Knowledge structures for integrating working and learning: A reflection on a decade of learning technology research for workplace learning. *Br. J. Educ. Technol.* **2019**, *51*, 331–346. [CrossRef]
4. Wille, R. Conceptual Landscapes of Knowledge: A Pragmatic Paradigm for Knowledge Processing. In *Classification in the Information Age, Proceedings of the 22nd Annual GfKl Conference, Dresden, Germany, 4–6 March 1998*; Gaul, W., Locarek-Junge, H., Eds.; Springer: Berlin/Heidelberg, Germany, 1999; pp. 344–356.
5. Wille, R. Restructuring Lattice Theory: An Approach Based on Hierarchies of Concepts. In *Ordered Sets*; Rival, I., Ed.; NATO Advanced Study Institutes Series; Springer: Dordrecht, The Netherlands, 1982; Volume 83, pp. 445–470. [CrossRef]
6. Ganter, B.; Wille, R. *Formal Concept Analysis—Mathematical Foundations*; Springer: Berlin/Heidelberg, Germany, 1999.
7. Hitzler, P.; Schärfe, H. (Eds.) *Conceptual Structures in Practice*; Chapman & Hall: London, UK, 2009.
8. Wille, R. Methods of Conceptual Knowledge Processing. In *Formal Concept Analysis*; Missaoui, R., Schmid, J., Eds.; Springer: Berlin/Heidelberg, Germany, 2006; Volume 3874, pp. 1–29.
9. Sacarea, C.; Zavaczki, R. Improving User's Experience in Navigating Concept Lattices: An Approach Based on Virtual Reality. In Proceedings of the 8th International Workshop "What can FCA do for Artificial Intelligence?" (FCA4AI 2020) Co-Located with 24th European Conference on Artificial Intelligence (ECAI 2020), Santiago de Compostela, Spain, 29 August 2020; Volume 2729, pp. 19–30.
10. Sacarea, C.; Sotropa, D.; Zavaczki, R. Toscana goes 3D: Using VR to Explore Life Tracks. In Proceedings of the Supplementary Proceedings ICFCA 2019 Conference and Workshops, Frankfurt, Germany, 25–28 June 2019; pp. 82–87.
11. Burmeister, P. *Formal Concept Analysis with ConImp: Introduction to the Basic Features*; Technische Universität Darmstadt: Darmstadt, Germany, 2003.
12. Cherukuri, A.K.; Jonnalagadda, A. Exploring Attributes with Domain Knowledge in Formal Concept Analysis. *J. Comput. Inf. Technol.* **2013**, *21*, 109–123. [CrossRef]
13. Kis, L.L.; Sacarea, C.; Troanca, D. FCA Tools Bundle—A Tool that Enables Dyadic and Triadic Conceptual Navigation. In Proceedings of the 5th International Workshop "What can FCA do for Artificial Intelligence"? Co-Located with the European Conference on Artificial Intelligence, FCA4AI@ECAI 2016, The Hague, The Netherlands, 30 August 2016; pp. 42–50.
14. Alam, M.; Le, T.N.N.; Napoli, A. Steps Towards Interactive Formal Concept Analysis with LatViz. In Proceedings of the 5th International Workshop "What can FCA do for Artificial Intelligence"? Co-Located with the European Conference on Artificial Intelligence, FCA4AI@ECAI 2016, The Hague, The Netherlands, 30 August 2016; pp. 51–62.
15. Robertson, G.; Mackinlay, J.; Card, S. Cone Trees: Animated 3D Visualizations of Hierarchical Information. In Proceedings of the SIGCHI Conference on Human Factors in Computing Systems, New Orleans, LA, USA, 27 April–2 May 1991; pp. 189–194. [CrossRef]
16. Steen, L.; Seebach, A. *Counterexamples in Topology*; Springer: Berlin/Heidelberg, Germany, 1970.
17. Engelking, R. *General Topology*; PWN: Warszawa, Poland, 1977.

Article

Influence of Highly Inflected Word Forms and Acoustic Background on the Robustness of Automatic Speech Recognition for Human–Computer Interaction

Andrej Zgank

Faculty of Electrical Engineering and Computer Science, University of Maribor, 2000 Maribor, Slovenia; andrej.zgank@um.si; Tel.: +386-2-220-7206

Abstract: Automatic speech recognition is essential for establishing natural communication with a human–computer interface. Speech recognition accuracy strongly depends on the complexity of language. Highly inflected word forms are a type of unit present in some languages. The acoustic background presents an additional important degradation factor influencing speech recognition accuracy. While the acoustic background has been studied extensively, the highly inflected word forms and their combined influence still present a major research challenge. Thus, a novel type of analysis is proposed, where a dedicated speech database comprised solely of highly inflected word forms is constructed and used for tests. Dedicated test sets with various acoustic backgrounds were generated and evaluated with the Slovenian UMB BN speech recognition system. The baseline word accuracy of 93.88% and 98.53% was reduced to as low as 23.58% and 15.14% for the various acoustic backgrounds. The analysis shows that the word accuracy degradation depends on and changes with the acoustic background type and level. The highly inflected word forms' test sets without background decreased word accuracy from 93.3% to only 63.3% in the worst case. The impact of highly inflected word forms on speech recognition accuracy was reduced with the increased levels of acoustic background and was, in these cases, similar to the non-highly inflected test sets. The results indicate that alternative methods in constructing speech databases, particularly for low-resourced Slovenian language, could be beneficial.

Keywords: human–computer interaction; automatic speech recognition; acoustic modeling; highly inflected word forms; acoustic background

MSC: 68T10

Citation: Zgank, A. Influence of Highly Inflected Word Forms and Acoustic Background on the Robustness of Automatic Speech Recognition for Human–Computer Interaction. *Mathematics* 2022, 10, 711. https://doi.org/10.3390/math10050711

Academic Editors: Grigoreta-Sofia Cojocar and Adriana-Mihaela Guran

Received: 30 December 2021
Accepted: 22 February 2022
Published: 24 February 2022

Publisher's Note: MDPI stays neutral with regard to jurisdictional claims in published maps and institutional affiliations.

Copyright: © 2022 by the author. Licensee MDPI, Basel, Switzerland. This article is an open access article distributed under the terms and conditions of the Creative Commons Attribution (CC BY) license (https:// creativecommons.org/licenses/by/ 4.0/).

1. Introduction

The convergence of Internet of Things (IoT) systems, services, and telecommunication networks has resulted in the omnipresence of human–computer interaction. Users can access services 24/7 by applying different devices. To support the human–computer interaction in a most natural way for users, procedures such as Automatic Speech Recognition (ASR) are needed [1]. A large vocabulary continuous speech recognition task can, on one side, be applied for standard human–computer interaction [2], or it can be used for producing text from various media material and user's content [3]. The text can be used for navigating and controlling [4,5], subtitling, indexing [6], translating, topic detection [7] ... Other examples of ASR applications besides human–computer interface input are the broadcast news speech recognition systems, massive open online courses, YouTube videos, and other various user content types, generated in intelligent ambient [8,9]. While an ASR can produce reasonable results in the case of typical conditions, the ASR's accuracy degrades significantly for adverse acoustic conditions [10,11] or when more complex languages are adopted [12,13]. Various acoustic backgrounds increased significantly in the last decade, where users interact with devices in different situations and record content in diverse environments. The omnipresent availability of smartphones in society has changed the role of who is recording and publishing the content.

The Slovenian University of Maribor (UMB) Broadcast News (BN) speech recognition system [14] is one such system for recognizing speech in a complex language with different acoustic backgrounds. It achieves approximately 15% worse overall speech recognition performance than similar systems for Western European languages. Possible causes are the linguistic characteristics of the Slovenian language, which is highly inflected and has a relatively free word order. An additional cause is the smaller amount of available Slovenian spoken language resources compared to other major languages like English, Mandarin, or German. While the research topic of under-resourced languages is covered broadly [12], there is far less attention given to the topic of highly inflected languages in automatic speech recognition. The preliminary analysis of achieved speech recognition results showed [15] that the worse performance belongs to the test set with the speech in the presence of an acoustical background, where the decrease can be even more than 40%. When similar English Broadcast News speech recognition systems were confronted with such degraded speech, the overall decrease of performance was in the region of 20% [16].

The decrease in the Slovenian speech recognition performance was the motivation to carry out a detailed analysis of how highly inflected word forms and acoustic background influence the speech recognition accuracy for human–computer interaction. We propose a novel analysis approach, where instead of using the general speech database, a separate, new type of speech database is constructed for tests. The new speech database is focused solely on highly inflected word forms originating from the same stem. This is needed to be able to perform a narrow analysis of specific language characteristics. The experiments are carried out with the Slovenian UMB Broadcast News speech recognition system. A particular emphasis is given to highly inflected word forms, where first, their effect on acoustic modeling is analyzed, and second, the combined influence of highly inflected word forms and acoustic background is studied. To the best of the author's knowledge, no similar research on the combined influence of acoustic background and highly inflected word forms on the level of acoustic modeling, as proposed in this paper, has been carried out. Moreover, this presents a first attempt to construct a focused, highly inflected speech database to assess this phenomenon of human–computer interaction.

Our view is that such analysis can produce a detailed insight into how strongly these conditions lower speech recognition accuracy. The research hypothesis is that combined influence of acoustic background and highly inflected word forms intensify the speech recognition accuracy decrease and that it is possible to identify approximate thresholds, where new digital signal processing techniques and dedicated algorithms for training acoustic models for highly inflected languages could be beneficial to reduce this degradation. The proposed novel analysis is carried out for the Slovenian language, but the same approach with a dedicated speech database could also be used for other morphologically complex languages. Additional anticipation is that the analysis results could point toward efficient construction of speech databases for complex low-resourced Slovenian language, where it is challenging to collect several 1000 h of transcribed speech, as it is today's standard for major languages.

The paper is organized as follows. First, the related work about the acoustic background and highly inflected languages for automatic speech recognition as part of human–computer interaction is presented. Different modules and functional aspects of an automatic speech recognition system, important for our research, are described in Section 3. The results and discussion are given in Section 4. The paper ends with the conclusions in Section 5.

2. Related Work

The robustness of speech recognition systems against the acoustic background is a mature research question, where the first research was coexistent with the fundamental automatic speech recognition research [17,18]. Despite the advancement of the automatic speech recognition research field, the question of robustness against the acoustic background, most frequently against noise, still presents a challenge, mainly in the area of developing new and improved methods of increasing the automatic speech recognition accuracy [19,20]. The question of robust automatic speech recognition can be addressed from two main viewpoints. The first one is focused on the reduction of acoustic background and improved feature extraction algorithms [21,22]. In the area of automatic speech recognition, Mel-frequency cepstral coefficients (MFCC) [23] preset a general feature extraction algorithm, which is used frequently. Different authors [24–26] proposed several other more complex feature extraction algorithms to improve the robustness. One of the frameworks, which includes a large number of various feature extraction algorithms is openSMILE [27], popular in the case of emotion recognition. More details about advanced feature extraction for speech recognition can be found in [28] The second robustness viewpoint concentrates on robust acoustic modeling approaches, where neural networks have mainly been used in recent years [29]. One of the frequently used solutions how to improve the robustness of neural networks regarding the acoustic background is data augmentation [30]. In this case, the original acoustic training data is extended with artificially generated copies, where the various acoustic backgrounds are added [31,32]. Moreover, other original audio data characteristics, like speed or pitch [33] can be altered to augment the training set and make it more robust to different conditions. The area of highly inflected languages for automatic speech recognition is a much less addressed topic. This is mainly due to the fact that the majority of currently commercially interesting languages do not belong to this category or are compensating for the limitations of a highly inflected language with a large amount of available spoken language resources. In the case of highly inflected languages, the language modeling question was given more importance [34], but the question of how highly inflected word forms influence acoustic modeling is a much less researched area. One of the first research works focused on the effects of highly inflected word forms for language modeling for automatic speech recognition was carried out for the Czech [35,36] and Slovenian languages [37]. The most frequently proposed solution for addressing the question of high inflection on the level of language modeling is to apply a non-word basic unit for modeling [38]. The idea behind this approach is to split the highly inflected word forms into shorter forms, which have a higher frequency in spoken language resources. In general, as a result of this approach, the out-of-vocabulary rate is usually reduced significantly, which might lead to improved speech recognition accuracy. The first disadvantage of this modeling approach is the reduced length of basic units, which increases acoustic confusability. The second disadvantage is the reduced modeling power of the language model, as the sequence of n-grams now estimates shorter linguistic parts of a sentence. Different types of basic units for language modeling and their combinations were proposed: Morphological [39,40], grammatical [41,42], stem-ending [43]. The proposed approaches improve the speech recognition accuracy to some extent, but the lag to major language performance can still be observed. In recent years, similar approaches used for highly inflected language modeling for automatic speech recognition were also adapted successfully to machine translation [38].

On the level of acoustic modeling for highly inflected automatic speech recognition, less research work was presented in the literature [44–46]. The shorter basic units from language modeling were incorporated directly into acoustic modeling, with necessary modifications to the decoding algorithms [47], which was used partly to compensate the reduced estimation power of the shorter n-gram units.

3. Materials and Methods

The performance of a speech recognition system depends largely on the acoustic characteristics of the input signal and on the properties of the supported language. Both conditions, as well as the experimental setup will be described below.

3.1. Acoustic Background and Speech Recognition

If the input signal $x(t)$ contains only speech signal $s(t)$ and acoustic background conditions are clean, high word accuracy can be expected, even for complex tasks such as continuous or spontaneous speech recognition [1]. In the case when the acoustic background signal $b(t)$ is present, the speech signal gets corrupted, and the word accuracy decreases. The corrupted input signal $x'(t)$ can be defined as in Equation (1):

$$x'(t) = s(t) + b(t) \tag{1}$$

The word accuracy decrease depends on the type and characteristics of the background signal $b(t)$ (e.g., noise, music, speech ...) and its overall energy and spectral characteristics in comparison with the main speech signal $s(t)$. The energy ratio between both signals can be estimated with the signal-to-noise ratio (SNR, Equation (2)), where the background signal is treated as noise:

$$\text{SNR}(\text{dB}) = 10 \cdot \log \frac{P_{speech}}{P_{background}} \tag{2}$$

where P_{speech} denotes the overall energy of the speech signal and $P_{background}$ denotes the overall energy of the background signal. The most important characteristic of background signal that influences the level of corruption is the spectrum, which varies, and is non-stationary.

In the detailed analysis presented in this paper, the acoustic models were trained on a speech database with various acoustic backgrounds. The baseline test set used only clean speech, while the dedicated test sets applied music as background, which corrupted the original clean speech signal. Music as a background source presents a realistic scenario, as such type of degradation occurs in the case of user content and input signals captured in human–computer interaction in intelligent ambient.

3.2. Acoustic Modeling and Highly Inflected Language

There are approximately 7000 live languages in the world, with 90% of them having less than 100,000 speakers. The human language technology support for various languages is important for their existence in the digital era, as it provides the core technology for natural human–computer interaction. Different languages can present a challenging task for a speech recognition system, due to their properties. Examples of such languages are tonal, highly inflected, or agglutinative languages. In the case of a highly inflected language, a single word stem is modified with many different suffixes or prefixes, which results in a large number of word forms. For some cases, the resulting number of word forms, modified from common stem, can exceed 100. As a consequence, an automatic speech recognizer's lexicon has to include a much larger number of words to achieve the same out of vocabulary rate as those speech recognition systems for a non-inflected language. It is crucial to achieving low out of vocabulary rate, as all the missing words from the vocabulary are being reflected directly in speech recognition errors as substitutions, with additional errors (approx. factor 1.5× might apply) coming from language modeling. Highly inflected language modeling approaches, such as stem-ending ones, can, to some extent, limit this effect of high out of vocabulary rate, but they lose a part of the n-gram prediction power, as shorter basic units span over shorter linguistic structures of a sentence.

Some of the languages belonging to the highly inflected group are Arabic, Slavic (e.g., Russian, Czech, Slovak, Polish, Slovenian, ...), Baltic (e.g., Lithuanian, and Latvian), and Indic languages. Examples of the Slovenian highly inflected word "hiša" (ENG: "house") in different cases for singular, dual, and plural, are given in Table 1.

Highly inflected word forms can be very similar acoustically, as can be seen from the example in Table 1, where only the last (or last few) phoneme of a relatively short word is modified. This acoustic confusability can to some degree be compensated with the n-grams of a statistical language model but can still present a severe problem for the speech decoder, especially in the case of the presence of acoustic background or accented speech. The Slovenian language has more than 50 dialects, which are spoken by approximately

2 million speakers. This results in a vast acoustic diversity, making the Slovenian language very suitable for our experiments of developing a dedicated speech database with highly inflected word forms. An additional challenge is given by the speaking style type, where two main categories are read/planned and spontaneous speech. In the case of spontaneous speech, an end vowel reduction often occurs in many Slovenian dialects, which additionally increases the acoustic confusability of highly inflected word forms. These characteristics clearly present the complexity of the automatic speech recognition task in the case of such languages.

Table 1. Example of Slovenian highly inflected word "hiša".

Singular		Dual		Plural	
Word	Suffix	Word	Suffix	Word	Suffix
hiša	-a	hiši	-i	hiše	-e
hiše	-e	hiš	/	hiš	/
hiši	-i	hišama	-ama	hišam	-am
hišo	-o	hiši	-i	hiše	-e
hiši	-i	hišah	-ah	hišah	-ah
hišo	-o	hišama	-ama	hišami	-ami

3.3. Speech Databases

The Slovenian BNSI Broadcast News speech database [48] belongs to the group of similar speech databases for Central and Eastern European languages [3,49], which were developed in the last two decades with the objective to provide suitable speech recognition resources for complex under-resourced languages. The BNSI speech database was used for research in the area of spoken language technologies and the design of the UMB BN speech recognition system. The Slovenian BNSI Broadcast News speech database is available via European Language Resources Association (ELRA/ELDA). Due to the specific requirements of the proposed analysis procedure, where only the acoustic characteristics should be considered, two additional speech databases were needed for the evaluation. The first one was the Slovenian SNABI SSQ Studio speech database, which was used for evaluating the general influence of acoustic background on continuous speech recognition accuracy. The Slovenian SNABI SSQ Studio speech database is available via Clarin SI. The second one was a dedicated test speech database, HI_SI, built for evaluating the influence of highly inflected word forms on acoustic modeling. The test scenarios, with isolated words, were selected for the evaluation, to avoid the influence of the language model, which could mask the acoustic differences which needed to be analyzed.

The Slovenian BNSI Broadcast News speech database consists of a total of 36 h of spoken material, where 30 h are used for acoustic training, 3 h for development, and the additional 3 h for evaluation. There are 1565 different speakers, 1069 of them are male and 477 are female. The gender of the remaining 19 speakers was unknown.

An important factor influencing speech recognition accuracy is acoustic background. There are several possible background types present [50] in a real-life environment: noise [10], speech, music. In a broadcast news speech database, the acoustic background is reflected in the parameter called acoustic conditions suitably simulating different conditions in which human–computer interaction needs to operate. The acoustic conditions and acoustic background in the BNSI speech database are presented in Table 2.

A large amount of the various acoustic conditions in the BNSI database is suitable for acoustic training, as it helps to increase the robustness of acoustic models against different acoustic conditions present in the evaluation set. However, it must also be taken into account that it could be difficult to train acoustic models for those acoustic conditions reliably with a relatively small proportion of the training set (e.g., F2, F5, FX).

Table 2. Acoustic conditions in the Slovenian BNSI Broadcast News speech database.

Acoustic Condition	Speech Type	Acoustic Background	Ratio (%)
F0	read	clean	36.56
F1	spontaneous	clean	16.23
F2	read/spontaneous	telephone channel	1.65
F3	read/spontaneous	music	6.02
F4	read/spontaneous	other	37.63
F5	nonnative	various	0.05
FX	other	other	1.86

The SNABI SSQ Studio test set used for the first part of the analysis consists of recordings of 37 speakers, where each speaker uttered different words in a silent studio environment. Thus, the baseline acoustic conditions of the test set were comparable to the ones present in the F0 and F1 acoustic condition parts of the training set. The first test set was based on isolated digits (12 words in the vocabulary, 441 recordings). The isolated digits' scenario is a non-complex one, but it is used frequently in different languages and applications, and thus enables a wide overall comparison of results. The second one, on city names (37 words in the vocabulary, 1360 recordings), is a more acoustically diverse test set, where the city names guarantee the absence of highly inflected word forms. This setup was used for baseline evaluation in the first part of the analysis. Dedicated test sets, where the baseline test set was combined with various acoustic backgrounds, were prepared for our experiment. The first type of acoustic background was instrumental music, similar to those frequently present in different media. The second type of acoustic background was based on songs, where instrumental music was combined with singing. This acoustic background type will be denoted in the remaining part of this paper as vocal music. Different types and SNR ratios of acoustic background generated for the dedicated SNABI test scenarios are presented in Table 3.

Table 3. Acoustic background in the dedicated SNABI test sets.

Acoustic Background Level	Instrumental Music SNR(dB)	Vocal Music SNR(dB)
Low	30.74	30.14
mid-low	22.30	22.06
mid	15.15	15.03
mid-high	10.77	10.69
high	5.10	5.03

The first part of the evaluation in Section 4 was carried out with the baseline clean and five acoustically degraded dedicated test sets. In the case of broadcast news recordings, the majority of the acoustic background has lower energy levels (i.e., higher SNR), and its type is usually instrumental.

The proposed HI_SI speech database was built for providing the dedicated test set with highly inflected word forms in a silent environment and in the presence of an acoustic background. During the preparation of experiments, several existing Slovenian speech databases were taken into account, but none of them comprised a test set suitable for evaluating acoustic confusability between highly inflected word forms. The HI_SI speech database contains recordings of 10 speakers, uttering various test sets/scenarios with six different highly inflected words generated from a single stem. In addition, one reference test scenario was recorded, which contains only words with a different stem in the basic word form. Three highly inflected test scenarios and the reference basic word form scenario were then selected as a baseline for the second evaluation part:

- H test set: Slovenian word "hiša" (Eng.: house), 6 inflected word forms,
- T test set: Slovenian word "telefon" (Eng.: telephone), 6 inflected word forms,
- M test set: Slovenian word "monitor" (Eng.: monitor), 6 inflected word forms,
- Reference test set: 6 acoustically diverse Slovenian words.

Instrumental music at three different levels was applied as an acoustic background for the second part of the evaluation. Different SNR ratios of acoustic background generated for the selected HI_SI test scenarios are presented in Table 4.

Table 4. Acoustic background in the dedicated HI_SI test sets.

Acoustic Background Level	H Test Set SNR (dB)	T Test Set SNR (dB)	M Test Set SNR (dB)	Reference Test Set SNR (dB)
low	31.18	30.64	30.96	31.02
mid	14.86	15.17	14.92	15.21
high	5.47	5.29	5.18	5.20

The acoustic properties of the HI_SI test sets, combined with the instrumental music as background, were comparable with the SNABI SSQ test sets, which is a favorable starting point for analyzing the influence of highly inflected word forms on acoustic modeling in the presence of an acoustic background.

The highly inflected test scenario was prepared in such a way that the city names isolated words' grammar was taken as a baseline and augmented with the highly inflected word forms from the particular HI_SI test set. Four additional isolated grammars were produced as an end result.

3.4. Experimental Setup

A general automatic speech recognition system is comprised of five basic building blocks (Figure 1; from left to right): Feature extraction front-end, acoustic models, language model, lexicon, and decoder. Front-end first pre-processes the input speech signal including the pre-emphasis and windowing and then extracts features. The extracted feature vectors are used as the input data for the speech decoder, which is applied to estimate the most probable hypothesis according to the models and feature vectors. There are three different types of models needed by a general speech decoder. The first are acoustic models, which represent the acoustic-phonetic parameters of acoustic material involved in the training process [50]. The language model is used to estimate the language characteristics in the form of a statistical model or grammar. It is built on a large text corpus in form of n-grams or using a grammatical description language in case of isolated words grammar. The last model is the phonetic vocabulary, which is used to transform the orthographic representation of words into phoneme form, which is generally needed for acoustic modeling and recognition. The output of a speech decoder is text result, which represents the word sequence uttered by the speaker in the input signal. A scheme of an automatic speech recognition system, with all the test sets included in the proposed evaluation, is given in Figure 1. The evaluation is divided into first and second part. The first one is focusing on various acoustic background, while the second one analyses the influence of highly inflected word forms with the proposed approach.

The experimental setup was designed in such a way that acoustic-phonetic properties played the most important role in the design and analysis process. The focus was given to the acoustic models, whose training procedure is described below. The automatic speech recognition system used for the analysis was built on three state left-right hidden Markov models (HMM), which were applied for acoustic modeling (Figure 2).

The limited amount of available Slovenian speech training data dictated the decision to use the HMM acoustic models instead of deep neural network (DNN) models. The first acoustic models' type usually performs more stable, particularly under challenging conditions.

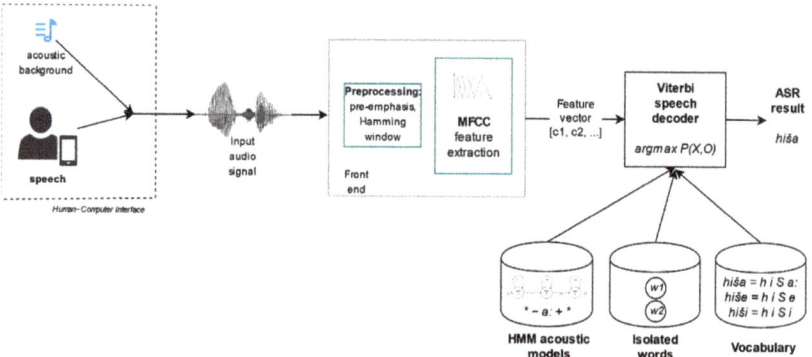

Figure 1. Scheme of the automatic speech recognition system used for the analysis.

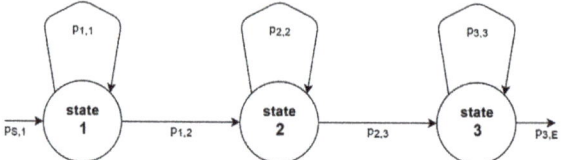

Figure 2. Three state left-right HMM acoustic model.

3.4.1. Preprocessing the Speech Signal

The captured speech signal s must be first prepared for acoustic training using the preprocessing methods. The pre-emphasis, as defined in Equation (3):

$$s'[n] = s[n] - 0.97 \cdot s[n-1] \qquad (3)$$

was applied to amplify the higher frequency band of the speech signal s with a high pass filter. Then, the signal s was split to produce frames, which were 25 ms long and were shifted with 10 ms step. A Hamming window function, defined in Equation (4):

$$s'[n] = \left\{ 0.54 - 0.46 \cos\left(\frac{2\pi(n-1)}{N-1}\right) \right\} \cdot s[n] \qquad (4)$$

was used to transform samples in each frame. N represented the length of Hamming window and n the particular sample.

3.4.2. Feature Extraction

After preprocessing, the feature extraction was done. Twelve Mel-scale frequency cepstrum coefficients (MFCC) were used as basic feature vectors. The MFCC were built around Mel-scale filter bank, which represented the nature of non-linear human speech perception and was linking the Mel (m) and frequency (f) dimension, as defined in Equation (5):

$$m = 2595 \cdot log_{10}\left(1 + \frac{f}{700}\right). \qquad (5)$$

The particular MFCC coefficient c_i was calculated using the Equation (6):

$$c_i = \sqrt{\frac{2}{N}} \sum_{j=1}^{N} m_j \cos\left(\frac{\pi i}{N}(j - 0.5)\right). \qquad (6)$$

As the 13th feature, the signal energy E was added to the vector, as defined in Equation (7):

$$E = \log \sum_{n=1}^{N} s^2[n] . \tag{7}$$

The basic set of 13 features was needed to calculate the first and second-order derivatives, which proved to model the short-term variations present in the speech signal. The final feature vector involved in the acoustic modeling contained 39 different coefficients extracted from the input speech signal.

3.4.3. The Acoustic Model and Training

The speech recognizer's acoustic HMM models applied a three-state topology with a weighted sum of continuous Gaussian probability density functions (PDF). The acoustic model λ was set as (Equation (8)):

$$\lambda = (A, B, \pi), \tag{8}$$

where A presented the state transition probabilities (a_{ij}), B the observation probability distribution (b_j) per each state j and π the initial values. For each observation o_t the probability b_j that it was generated, was defined as (Equation (9)):

$$b_j(o_t) = \sum_{m=1}^{M} c_{jm} \mathcal{N}(o_t ; \mu_{jm}, \Sigma_{jm}), \tag{9}$$

where c_{jm} was the weight of each mixture, M the total number of mixtures and $\mathcal{N}$ the Gaussian PDF with mean vector μ and diagonal covariance matrix Σ (Equation (10)):

$$\mathcal{N}(o ; \mu, \Sigma) = \frac{1}{\sqrt{(2\pi)^n |\Sigma|}} e^{-\frac{1}{2}(o-\mu)' \Sigma^{-1}(o-\mu)} \tag{10}$$

The acoustic models were speaker and gender independent, with graphemes as basic acoustic units. Graphemes were used instead of phonemes as basic units as they reduce the influence of phonetic vocabulary errors which result from the complex grapheme-to-phoneme task for the Slovenian language. The HMM acoustic models were trained using the Baum-Welch re-estimation, which is defined in the form of forward probability α_j (Equation (11)) using the forward recursion:

$$\alpha_j(t) = \left[\sum_{i=2}^{N-1} \alpha_i(t-1) a_{ij} \right] b_j(v_t), \tag{11}$$

where $\alpha_1(1) = 1$ and $\alpha_j(1) = a_{1j} b_j(o_1)$ are set as initial conditions and j is $1 < j < N$. The final step is defined as (Equation (12)):

$$\alpha_N(T) = \left[\sum_{i=2}^{N-1} \alpha_i(T) a_{iN} \right] \tag{12}$$

The backward probability $\beta_j(t)$ is defined by Equation (13) using the backward recursion:

$$\beta_i(t) = \sum_{j=2}^{N-1} a_{ij} b_j(o_{t+1}) \beta_j(t+1), \tag{13}$$

where i and t are $1 < i < N$ and $T > t > 1$. The initial value is set as $\beta_i(T) = a_{iN}$ with the final step defined as (Equation (14)):

$$\beta_1(1) = \sum_{j=2}^{N-1} a_{1j} b_j(o_1) \beta_j(1). \tag{14}$$

3.4.4. Training of Context-Independent Acoustic Models

The above-defined algorithms were used for training the acoustic models in three steps. The first two used context-independent modeling. The last one also modeled the context of a single acoustic unit, which can improve the acoustic modeling performance significantly [51]. This approach was applied to improve the quality of speech database transcriptions stepwise. The original transcriptions were produced manually during the speech database development but contained some errors.

First, the context-independent acoustic models' initial parameters were set as global values, equal for all different models. The initial values were estimated using the Baum-Welch re-estimation. For acoustic training, a small subset was randomly selected from the full training set. Three iterations of the training procedure were carried out. The results were acoustic models with 1 Gaussian PDF per state. Next, the forced realigning of the speech database's transcriptions was used to detect the outliers in the training set.

In the second step, the context-independent acoustic models were initialized from the scratch. The initial acoustic models with 1 Gaussian PDF were trained in a stepwise manner until reaching the mixture of 32 Gaussian PDF per state. Such a type of acoustic model has a better generalization effect. They are more suitable for the evaluation of an unseen test set comprising different acoustic conditions. These acoustic models were again applied for the forced-realigning procedure. The result was an improved version of training set speech transcriptions.

3.4.5. Training of Context-Dependent Acoustic Models

The third step started again with the initialization of acoustic models. This time local initial values for each acoustic unit were used. The cross-word context-dependent (trigrapheme) acoustic models were created after seven training iterations. The context-dependent acoustic models improve the performance in presence of the coarticulation effect [52]. The context-dependent acoustic models have a large number of free parameters. These were reduced with the use of a decision tree-based clustering algorithm. Its objective was to tie together all similar acoustic models' states. A data-driven approach [53] based on the acoustic models' confusion matrix was used to produce broad classes. In the end, the number of Gaussian PDF per state was increased to 16. These were the final acoustic models, which were applied to the speech recognition task. For more details on the UMB BN speech recognition system see [14,15].

4. Results and Discussion

The first part of the analysis was focused on the influence of acoustic background on speech recognition accuracy. The experiments were carried out in several steps using the UMB BN speech recognition system and dedicated test sets generated from the SNABI SSQ Studio database. The speech recognition results for isolated words recognition are presented in the form of Word Accuracy (WA), which is defined as in Equation (15):

$$Word\ accuracy(\%) = \frac{H-I}{N} \cdot 100 \qquad (15)$$

where H denotes the number of correctly recognized words in the test set, I is the number of insertions and N denotes the number of all words in the test set.

For comparison reasons, a short overview of previous general speech recognition results with the UMB BN large vocabulary continuous speech recognition system will be given. The comparable setup of the UMB BN system, with the BNSI Broadcast News speech database [4], achieves a 73.26% word accuracy on the reference BNSI evaluation set, which includes all acoustic conditions. In the case of clean read studio speech [14], the word accuracy was 84.81%, but it dropped to just 63.88% when various acoustic backgrounds (F3, F4, and Fx conditions) were also present. These results can be directly compared with the Slovenian ASR system developed by Golik et al. [54], who has used the identical BNSI Broadcast News speech database, but different system architecture achieved 72.2% word accuracy on the evaluation set. Additional comparisons can be made with broadcast news speech recognition systems for other languages or with cross-lingual setups [55,56]. The broadcast news ASR system for Slovak, a similar Slavic language, presented by Viszlay et al. [57], performed with 78.04% word accuracy on a more extensive training speech database. The Slovenian ASR system [58] built with cross-lingual bootstrapped acoustic models and retrained on lightly supervised public data achieved 78.51% word accuracy. The comparison with published results shows that the UMB BN large vocabulary continuous speech recognition system has similar performance and can be used to carry out the proposed analysis.

First, the analysis was devoted to evaluating both baseline SNABI test sets without acoustic background with the UMB Broadcast News speech recognition system baseline. Word accuracy for isolated digits was 93.88% and for the city names 98.53%. Although the city names test set includes more words in the grammar than the isolated digits' test set, it achieved better performance, due mainly to the longer words. In the case of isolated digits, the average word length was 3.7 graphemes, and for the city names the average word length was 5.2 graphemes. Longer words are acoustically diverse and thus give the speech recognition decoder more data to estimate the most probable grapheme sequence.

In the second step of the first part of the analysis, the SNABI isolated digits and city names' test sets with mixed acoustic backgrounds were tested with the UMB Broadcast News speech recognition system. The first test set applied instrumental music for the acoustic background and the second one used vocal music. The speech recognition results for five different acoustic background levels per test set are given in Tables 5 and 6.

Table 5. The isolated digits' speech recognition results in the presence of various acoustic backgrounds.

Test Set	Acoustic Background Level	Instrumental Music WA (%)	Vocal Music WA (%)
isolated digits	clean	93.88	93.88
isolated digits	low	88.21	84.35
isolated digits	mid-low	69.16	65.08
isolated digits	mid	42.86	42.86
isolated digits	mid-high	29.25	33.56
isolated digits	high	23.58	28.12
isolated digits	average background level	50.61	50.80

Table 6. The city names speech recognition results in the presence of various acoustic backgrounds.

Test Set	Acoustic Background Level	Instrumental Music WA (%)	Vocal Music WA (%)
city names	clean	98.53	98.53
city names	low	94.12	89.93
city names	mid-low	65.39	62.89
city names	mid	32.70	36.44
city names	mid-high	20.21	24.98
city names	high	15.14	19.03
city names	average background level	45.51	46.65

When isolated digits mixed with acoustic background were recognized, the instrumental music decreased word accuracy from 93.88% to 23.58%. In the case of low and mid-low acoustic background levels, word accuracy was 88.21% and 69.16%, respectively. In the case when the vocal music was used for the acoustic background, the word accuracy reduced from 93.88% to 28.12%. Word accuracy of 84.35% and 65.08% was achieved with the low and mid-low acoustic background levels. The average word accuracy with instrumental music as a background was 50.61%, and 50.80% with vocal music as a background.

These results show that the type of acoustic background does have an influence on speech recognition accuracy. In the case of lower background level, the vocal music produced higher degradation than the instrumental music. This can be explained by the effect where the vocal part of music decreases the robustness of acoustic models, due to its similarity with ordinary speech found in uttered sentences.

An instrumental music background decreased the word accuracy of the city names test set from 98.53% to 15.14%, respectively. At low and mid-low background levels, the word accuracy was 94.12% and 65.39%, respectively. The vocal music as acoustic background degraded the word accuracy from 98.53% to 19.03%, respectively. The city names' average word accuracy was 45.51% for instrumental music background and 46.65% for the vocal music background. In the case of the city names test scenario, the difference in average word accuracy between both types of the acoustic background was slightly higher than for the previous experiment with isolated digits.

A detailed comparison of all four various test scenarios included in the first part of the analysis is given in Figure 3. The comparison between test scenarios confirms that the influence of acoustic background level also depends on the type and complexity of the test scenario. In the case of the city names test scenario, which has more words in the speech recognizer's vocabulary than the isolated digits scenario, the word accuracy decrease is smaller for low-level acoustic background (low condition), but the word accuracy degradation increased for higher levels of acoustic background (mid-high and high conditions). A plausible explanation is that the acoustic background increased the acoustic confusability between words in the speech recognizer's search space significantly.

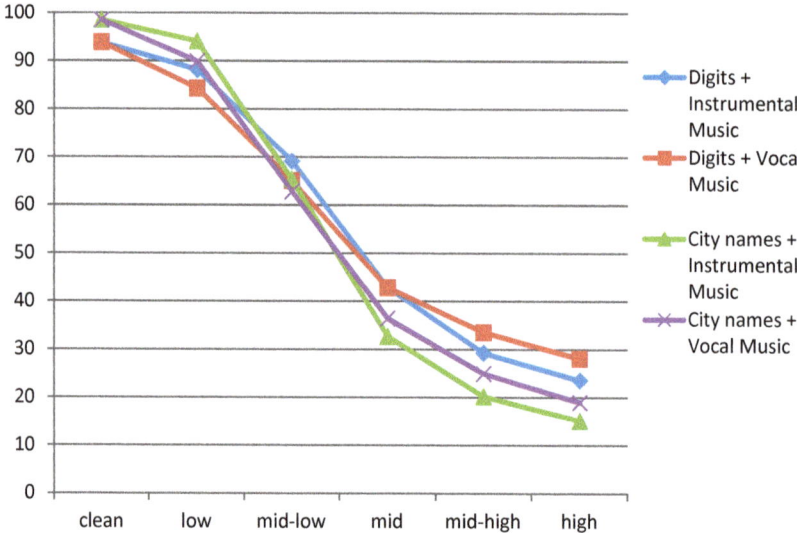

Figure 3. Comparison of different acoustic background levels on the speech recognizer's word accuracy.

The influence of acoustic background type on speech recognition accuracy swaps with the level of acoustic background. While for lower acoustic background levels vocal music degraded speech recognition performance in a more severe way, the situation changed after introduction of the medium level of acoustic background. In the case of higher acoustic background level, the instrumental music showed higher degrading factors on speech recognition accuracy. A possible reason is that vocals in music at the low level of acoustic background present a sort of background speech, which is known to decrease the word accuracy significantly. With the increased level of acoustic background, this effect of background speech diminished, and the degradation of musical acoustic background as a whole began to be the most important influence factor.

The second part of the analysis was focused on speech recognition accuracy for highly inflected words. Utterances without acoustic background were tested first, with the goal to define the baseline. In the next step, recordings of highly inflected words were evaluated, with various levels of acoustic background (Table 7).

Table 7. Speech recognition results for the highly inflected HI_SI test sets.

Acoustic Background Level	H Test Set WA (%)	T Test Set WA (%)	M Test Set WA (%)	Non-Highly Inflected Test Set WA (%)
clean	71.7	63.3	80.0	93.3
low	76.7	60.0	76.7	93.3
mid	40.0	28.3	43.3	41.7
high	15.0	11.7	13.3	13.3
average	43.9	33.3	44.4	49.4

The reference test set (Table 7) comprised only non-highly inflected words and city names and was used as a control case, showing how the acoustic background influenced speech recognition accuracy for this specific scenario. The achieved results (WA between 93.3% and 13.3%) with this set were comparable to the ones in the first part of the analysis. This confirms that the newly proposed highly inflected HI_SI speech database provides comparable data for evaluation. The highly inflected test sets produced significantly worse results when tested in clean acoustic conditions. The word accuracy dropped to 71.7%, 63.3%, and 80.0% for the H, T, and M test sets, respectively. The detailed analysis of speech recognition results revealed that, in the case of a clean acoustic background, all speech recognition errors occurred only between the highly inflected word forms (e.g., uttered word form "monitorja" was, for example, substituted with the form "monitorju"). The most frequent highly inflected word form substitution occurred in the H test set where, in 47.1% of error cases, the uttered word form "hiša" was misrecognized as word form "hišah". In the case of both highly inflected words, the difference in suffix is only minimal—the first suffix "-a" is acoustically very similar to the second suffix "-ah". This ending phoneme /h/ can frequently be reduced in various Slovenian accents. Moreover, the phoneme /h/ acoustic characteristics (spectral energy, time variance) can present a difficult task for a speech recognition system. This type of substitution errors and a significant decrease in word accuracy point-out clearly the complexity of speech recognition for highly inflected languages. When a low-level acoustic background was added to the highly inflected HI_SI test sets T and M, additional word accuracy degradation was observed—60.0% and 76.7%, respectively. The first substitutions to the city names were observed in this case. This shows that the acoustic background increased the acoustic confusability additionally, beyond the point of modified word endings. For the Reference (93.3%) and H test set (76.7%), the accuracy was preserved or even improved. The possible cause for this is the statistical approach used for acoustic modeling. The decrease of word accuracy between the Reference test set and highly inflected H, T, and M, test sets was still comparable with the clean acoustic background configuration. With the increased level of acoustic background (mid and high levels), the word accuracy decreased statistically significantly for all four test sets similarly. The effect of acoustic confusability of highly inflected word forms was now minimized, as the higher level of the acoustic background masked the speech signal and its characteristics important for speech recognition algorithms. This was also confirmed with the detailed analysis of speech recognition results, where the substitutions were almost evenly distributed between errors in highly inflected word forms and city names.

The overall results show that, in the most frequent speech recognizers' operating conditions, the additional absolute word accuracy degradation between 15% and 25% caused by acoustic modeling can be expected for highly inflected languages. The experimental results confirmed the research hypothesis. A specific area above 15 dB SNR was identified from results where it would be of particular benefit to apply advanced acoustic modeling methods for highly inflected languages to improve the speech recognition results. Below the 15 dB SNR level, the influence of highly inflected word forms on speech recognition accuracy diminishes, and the general influence of acoustic background comes in front.

5. Conclusions

The proposed analysis approach showed the severe impact of acoustic background on speech recognition accuracy. While acoustic background at low energy decreased the word accuracy in a limited way, the high energy of acoustic background presented a limitation for using an ASR. One of the possible solutions would be to treat both cases separately, using different background reduction/decomposition methods.

The speech recognition analysis on highly inflected test scenarios showed a statistically significant additional impact of this language characteristic on word accuracy, particularly for none or a low level of acoustic backgrounds. This confirms that some languages are more challenging for automatic speech recognition tasks.

It would be interesting to extend the proposed analysis to other languages to check if similar estimations can also be observed for them. The prerequisite for such a task would be the availability of a dedicated highly inflected acoustic test set in the analyzed language. Currently, the well-established approach in acoustic modeling for automatic speech recognition is to collect as many hours of transcribed speech material as possible. This can result in several 10,000-h speech databases for major languages. However, this approach is unrealistic for many low-resource languages with a limited number of speakers and limited economic interest. The results of this analysis could in the future pave the way for an alternative acoustic modeling approach where the speech database would be designed in a way to better cover such language properties as are highly inflected word forms. This could reduce the need for collecting large amounts of transcribed speech material and ease the development.

Funding: The Slovenian Research Agency partially funded this research under Contract Number P2-0069 "Advanced methods of interaction in Telecommunication".

Institutional Review Board Statement: Not applicable.

Informed Consent Statement: Not applicable.

Data Availability Statement: The Slovenian SNABI SSQ Studio speech database is available via Clarin SI, http://hdl.handle.net/11356/1051 (accessed on 29 December 2021). The Slovenian BNSI Broadcast News speech database is available via ELRA/ELDA, http://www.islrn.org/resources/502-280-144-938-4/ (accessed on 29 December 2021).

Acknowledgments: The author thanks all those who contributed their speech recordings to the spoken language's resources used in these experiments.

Conflicts of Interest: The author declares no conflict of interest. The funders had no role in the design of the study; in the collection, analyses, or interpretation of data; in the writing of the manuscript, or in the decision to publish the results.

References

1. Lee, C.H. On Automatic Speech Recognition at the Dawn of the 21st Century. *IEICE Trans. Inf. Syst.* **2003**, *E86-D*, 377–396.
2. Maskeliunas, R.; Ratkevicius, K.; Rudzionis, V. Voice-based Human-Machine Interaction Modeling for Automated Information Services. *Electron. Electr. Eng.* **2011**, *110*, 109–112. [CrossRef]
3. Pleva, M.; Juhar, J. Building of Broadcast News Database for Evaluation of the Automated Subtitling Service. *Commun.-Sci. Lett. Univ. Zilina* **2013**, *15*, 124–128. [CrossRef]
4. Mięsikowska, M. Discriminant Analysis of Voice Commands in the Presence of an Unmanned Aerial Vehicle. *Information* **2021**, *12*, 23. [CrossRef]
5. Valizada, A.; Akhundova, N.; Rustamov, S. Development of Speech Recognition Systems in Emergency Call Centers. *Symmetry* **2021**, *13*, 634. [CrossRef]

6. Szaszak, G.; Tundik, A.M.; Vicsi, K. Automatic speech to text transformation of spontaneous job interviews on the HuComTech database. In Proceedings of the 2011 2nd International Conference on Cognitive Infocommunications (CogInfoCom), Budapest, Hungary, 7–9 July 2011.
7. Zlacky, D.; Staš, J.; Juhar, J.; Čižmar, A. Term weighting schemes for Slovak text document clustering. *J. Electr. Electron. Eng.* **2013**, *6*, 163–166.
8. Gondi, S.; Pratap, V. Performance Evaluation of Offline Speech Recognition on Edge Devices. *Electronics* **2021**, *10*, 2697. [CrossRef]
9. Beňo, L.; Pribiš, R.; Drahoš, P. Edge Container for Speech Recognition. *Electronics* **2021**, *10*, 2420. [CrossRef]
10. Pervaiz, A.; Hussain, F.; Israr, H.; Tahir, M.A.; Raja, F.R.; Baloch, N.K.; Ishmanov, F.; Bin Zikria, Y. Incorporating Noise Robustness in Speech Command Recognition by Noise Augmentation of Training Data. *Sensors* **2020**, *20*, 2326. [CrossRef]
11. Gnanamanickam, J.; Natarajan, Y.; Sri, S.P.K. A Hybrid Speech Enhancement Algorithm for Voice Assistance Application. *Sensors* **2021**, *21*, 7025. [CrossRef]
12. Besacier, L.; Barnard, E.; Karpov, A.; Schultz, T. Automatic speech recognition for under-resourced languages: A survey. *Speech Commun.* **2014**, *56*, 85–100. [CrossRef]
13. Wołk, K.; Wołk, A.; Wnuk, D.; Grześ, T.; Skubis, I. Survey on dialogue systems including slavic languages. *Neurocomputing* **2021**, *477*, 62–84. [CrossRef]
14. Maučec, M.S.; Žgank, A. Speech recognition system of Slovenian broadcast news. In *Speech Technologies*; InTech: Rijeka, Croatia, 2011; pp. 221–236.
15. Gank, A.; Donaj, G.; Maučec, M.S. UMB Broadcast News 2014 continuous speech recognition system: What is the influence of language resources' size? Language technologies. In Proceedings of the 17th International Multiconference Information Society—IS 2014, Ljubljana, Slovenia, 9–10 October 2014; Volume G.
16. Raj, B.; Parikh, V.; Stern, R. The effects of background music on speech recognition accuracy. In Proceedings of the 1997 IEEE International Conference on Acoustics, Speech, and Signal Processing, Munich, Germany, 21–24 April 1997; pp. 851–854. [CrossRef]
17. Gong, Y. Speech recognition in noisy environments: A survey. *Speech Commun.* **1995**, *16*, 261–291. [CrossRef]
18. Juang, B. Speech recognition in adverse environments. *Comput. Speech Lang.* **1991**, *5*, 275–294. [CrossRef]
19. Zhang, Z.; Geiger, J.; Pohjalainen, J.; Jin, W.; Schuller, B. Deep learning for environmentally robust speech recognition: An overview of recent developments. *ACM Trans. Intell. Syst. Technol.* **2018**, *9*, 1–28. [CrossRef]
20. Li, J.; Deng, L.; Gong, Y.; Haeb-Umbach, R. An Overview of Noise-Robust Automatic Speech Recognition. *IEEE/ACM Trans. Audio Speech Lang. Proc.* **2014**, *22*, 745–777. [CrossRef]
21. Upadhyay, N.; Rosales, H.G. Robust Recognition of English Speech in Noisy Environments Using Frequency Warped Signal Processing. *Natl. Acad. Sci. Lett.* **2018**, *41*, 15–22. [CrossRef]
22. Kang, B.O.; Jeon, H.B.; Park, J.G. Speech Recognition for Task Domains with Sparse Matched Training Data. *Appl. Sci.* **2020**, *10*, 6155. [CrossRef]
23. Zheng, F.; Zhang, G.; Song, Z. Comparison of different implementations of MFCC. *J. Comput. Sci. Technol.* **2001**, *16*, 582–589. [CrossRef]
24. Nassif, A.B.; Shahin, I.; Attili, I.; Azzeh, M.; Shaalan, K. Speech Recognition Using Deep Neural Networks: A Systematic Review. *IEEE Access* **2019**, *7*, 19143–19165. [CrossRef]
25. Raj, B.; Stern, R. Missing-feature approaches in speech recognition. *IEEE Signal Process. Mag.* **2005**, *22*, 101–116. [CrossRef]
26. Gupta, K.; Gupta, D. An analysis on LPC, RASTA and MFCC techniques in Automatic Speech recognition system. In Proceedings of the 2016 6th International Conference—Cloud System and Big Data Engineering (Confluence), Noida, India, 14–15 January 2016; pp. 493–497.
27. Eyben, F.; Wöllmer, M.; Schuller, B. Opensmile: The munich versatile and fast open-source audio feature extractor. In Proceedings of the 18th ACM International Conference on Multimedia, Firenze, Italy, 25–29 October 2010; pp. 1459–1462.
28. Anusuya, M.A.; Katti, S.K. Front end analysis of speech recognition: A review. *Int. J. Speech Technol.* **2011**, *14*, 99–145. [CrossRef]
29. Lee, K.H.; Kang, W.H.; Kang, T.G.; Kim, N.S. Integrated DNN-based model adaptation technique for noise-robust speech recognition. In Proceedings of the 2017 IEEE International Conference on Acoustics, Speech and Signal Processing (ICASSP), New Orleans, LA, USA, 5–9 March 2017; pp. 5245–5249. [CrossRef]
30. Ko, T.; Peddinti, V.; Povey, D.; Khudanpur, S. Audio augmentation for speech recognition. In Proceedings of the Sixteenth Annual Conference of the International Speech Communication Association—Interspeech 2015, Dresden, Germany, 6–10 September 2015.
31. Nguyen, T.-S.; Stuker, S.; Niehues, J.; Waibel, A. Improving Sequence-To-Sequence Speech Recognition Training with On-The-Fly Data Augmentation. In Proceedings of the ICASSP 2020—2020 IEEE International Conference on Acoustics, Speech and Signal Processing (ICASSP), Barcelona, Spain, 4–8 May 2020; pp. 7689–7693. [CrossRef]
32. Prisyach, T.; Mendelev, V.; Ubskiy, D. Data Augmentation for Training of Noise Robust Acoustic Models. In *International Conference on Analysis of Images, Social Networks and Texts*; Springer: Cham, Switzerland, 2016; pp. 17–25. [CrossRef]
33. Shahnawazuddin, S.; Adiga, N.; Kathania, H.K.; Sai, B.T. Creating speaker independent asr system through prosody modi-fication based data augmentation. *Pattern Recognit. Lett.* **2020**, *131*, 213–218. [CrossRef]
34. Staš, J.; Hladek, D.; Pleva, M.; Juhar, J. Slovak language model from Internet text data. In *Towards Autonomous, Adaptive, and Context-Aware Multimodal Interfaces: Theoretical and Practical Issues, LNCS 6456*; Springer: Berlin/Heidelberg, Germany, 2011; pp. 352–358.

35. Byrne, W.; Hajič, J.; Ircing, P.; Jelinek, F.; Khudanpur, S.; McDonough, J.; Peterek, N.; Psutka, J. Large Vocabulary Speech Recognition for Read and Broadcast Czech. In Proceedings of the Text, Speech and Dialogue—Second International Workshop, TSD'99, Plzen, Czech Republic, 13–17 September 1999; pp. 235–240. [CrossRef]
36. Ircing, P.; Krbec, P.; Hajic, J.; Psutka, J.; Khudanpur, S.; Jelinek, F.; Byrne, W. On large vocabulary continuous speech recognition of highly inflectional language-Czech. In Proceedings of the Seventh European Conference on Speech Communication and Technology, Aalborg, Denmark, 3–7 September 2001.
37. Maucec, M.S.; Kacic, Z.; Horvat, B. A framework for language model adaptation for highly-inflected Slovenian language. In *ISCA Tutorial and Research Workshop (ITRW) on Adaptation Methods for Speech Recognition*; ISCA: Sophia Antipolis, France, 2001.
38. Schwenk, H. Trends and challenges in language modeling for speech recognition and machine translation. In Proceedings of the 2009 IEEE Workshop on Automatic Speech Recognition & Understanding, Moreno, Italy, 13 November–17 December 2009; p. 23. [CrossRef]
39. Mousa, A.E.-D.; Shaik, M.A.B.; Schlüter, R.; Ney, H. Morpheme level hierarchical pitman-yor class-based language models for LVCSR of morphologically rich languages. In Proceedings of the Annual Conference of the International Speech Communication Association, Lyon, France, 25–29 August 2013. [CrossRef]
40. Staš, J.; Hladek, D.; Juhar, J. Morphologically motivated language modeling for Slovak continuous speech recognition. *J. Electr. Electron. Eng.* **2012**, *5*, 233–237.
41. Donaj, G.; Kačič, Z. Context-dependent factored language models. *EURASIP J. Audio, Speech, Music Process.* **2017**, *2017*, 6. [CrossRef]
42. Vazhenina, D.; Markov, K. Factored language modeling for Russian LVCSR. In Proceedings of the International Joint Conference on Awareness Science and Technology and Ubi-Media Computing, iCAST 2013 and UMEDIA 2013, Aizu-Wakamatsu, Japan, 2–4 November 2013.
43. Maucec, M.S.; Rotovnik, T.; Zemljak, M. Modelling Highly Inflected Slovenian Language. *Int. J. Speech Technol.* **2003**, *6*, 245–257. [CrossRef]
44. Karpov, A.; Kipyatkova, I.; Ronzhin, A. Very large vocabulary ASR for spoken Russian with syntactic and morphemic analysis. In Proceedings of the Annual Conference of the International Speech Communication Association, Florence, Italy, 27–31 August 2011. [CrossRef]
45. Pipiras, L.; Maskeliūnas, R.; Damaševičius, R. Lithuanian Speech Recognition Using Purely Phonetic Deep Learning. *Computers* **2019**, *8*, 76. [CrossRef]
46. Polat, H.; Oyucu, S. Building a Speech and Text Corpus of Turkish: Large Corpus Collection with Initial Speech Recognition Results. *Symmetry* **2020**, *12*, 290. [CrossRef]
47. Rotovnik, T.; Maučec, M.S.; Kačič, Z. Large vocabulary continuous speech recognition of an inflected language using stems and endings. *Speech Commun.* **2007**, *49*, 437–452. [CrossRef]
48. Zgank, A.; Verdonik, D.; Markus, A.Z.; Kacic, Z. BNSI Slovenian broadcast news database—Speech and text corpus. In Proceedings of the Eurospeech, 9th European Conference on Speech Communication and Technology, Lisbon, Portugal, 4–8 September 2005. [CrossRef]
49. Vicsi, K.; Szaszák, G. Using prosody to improve automatic speech recognition. *Speech Commun.* **2010**, *52*, 413–426. [CrossRef]
50. Bang, J.-U.; Kim, S.-H.; Kwon, O.-W. Acoustic Data-Driven Subword Units Obtained through Segment Embedding and Clustering for Spontaneous Speech Recognition. *Appl. Sci.* **2020**, *10*, 2079. [CrossRef]
51. Theera-Umpon, N.; Chansareewittaya, S.; Auephanwiriyakul, S. Phoneme and tonal accent recognition for Thai speech. *Expert Syst. Appl.* **2011**, *38*, 13254–13259. [CrossRef]
52. Verdonik, D. Between understanding and misunderstanding. *J. Pragmat.* **2010**, *42*, 1364–1379. [CrossRef]
53. Lopes, C.; Perdigao, F. Broad phonetic class definition driven by phone confusions. *EURASIP J. Adv. Signal Process.* **2012**, *2012*, 158. [CrossRef]
54. Golik, P.; Tüske, Z.; Schlüter, R.; Ney, H. Development of the RWTH transcription system for slovenian. In Proceedings of the 14th Annual Conference of the International Speech Communication Association, Lyon, France, 25–29 August 2013. [CrossRef]
55. Pleva, M.; Čižmar, A.; Juhar, J.; Ondaš, S.; Mirilovič, M. Towards Slovak Broadcast News Automatic Recording and Tran-scribing Service. In *Verbal and Nonverbal Features of Human-Human and Human-Machine Interaction, Lecture Notes in Computer Science 5042*; Springer: Berlin/Heidelberg, Germany, 2008; pp. 158–168.
56. Prochazka, V.; Pollak, P.; Zdansky, J.; Nouza, J. Performance of Czech Speech Recognition with Language Models Created from Public Resources. *Radio Eng.* **2011**, *20*, 1002–1008.
57. Viszlay, P.; Staš, J.; Koctúr, T.; Lojka, M.; Juhár, J. An extension of the Slovak broadcast news corpus based on semi-automatic annotation. In Proceedings of the Tenth International Conference on Language Resources and Evaluation—LREC 2016, Portorož, Slovenia, 23–28 May 2016; pp. 4684–4687.
58. Nouza, J.; Safarik, R.; Cerva, P. ASR for South Slavic Languages Developed in Almost Automated Way. In Proceedings of the 17th Annual Conference of the International Speech Communication Association, San Francisco, CA, USA, 8–12 September 2016. [CrossRef]

MDPI
St. Alban-Anlage 66
4052 Basel
Switzerland
Tel. +41 61 683 77 34
Fax +41 61 302 89 18
www.mdpi.com

Mathematics Editorial Office
E-mail: mathematics@mdpi.com
www.mdpi.com/journal/mathematics

www.ingramcontent.com/pod-product-compliance
Lightning Source LLC
LaVergne TN
LVHW070711100526
838202LV00013B/1069